EXPERIMENTS IN BASIC CHEMISTRY

FOURTH EDITION

Steven Murov
Brian Stedjee
Modesto Junior College

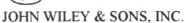

JOHN WILEY & SONS, INC.
New York • Chichester • Brisbane • Toronto • Singapore • Weinheim

ISBN 0-471-16030-X

Printed in the United States of America

10 9 8 7 6 5 4 3

Printed and bound by Bradford & Bigelow, Inc.

PREFACE

Our goal during the preparation of this laboratory text has been to provide a safe, interesting and challenging educational experience for students. Justifiable public concern for the environment and for student safety and soaring chemical disposal costs has placed ever increasing pressure on chemistry departments to provide improved conditions for undergraduate chemical instruction. We believe that this is one of the very few chemistry lab texts that avoids the use of known carcinogens and toxic metal salts such as nickel(II), lead(II), chromium(VI) and barium compounds. In addition, only one experiment employs corrosives more concentrated than 6 M. Experiments have been selected to maximize educational impact while minimizing hazards and cost. However, regardless of the precautions we have taken, it must be remembered that it is impossible to remove every potential danger from the chemical laboratory. Students should follow directions, safety rules should be adhered to carefully and no unauthorized experiment should be performed. Because safety concerns always should be first and foremost in the mind of the experimenter, the first (*page 4*) and last (*page 380*) exercises for students in the book involve safety considerations.

Whenever possible, an exploratory approach is used that encourages students to inquire and discover concepts as they experiment. This text has been designed for use by instructors who intend to be involved in the laboratory. It will often be necessary to explain and clarify concepts and to provide subtle details that facilitate the performance of the experiment. We have attempted to include useful hints in the instructor's manual. It is somewhat paradoxical that instructors teach the unit conversion method for solving problems in the lecture portion of the course and revert to a step-by-step method with real data in the laboratory. In this text, students are usually asked to calculate the answers using both approaches.

The principal changes in the 4th edition of this text involve the *Prelaboratory Exercises*, the sequence of the experiments and exercises, and additions to the *Exercise* section. The *Prelaboratory Exercises* have been printed on isolated pages so that they can be torn out and handed in without disturbing the *Procedure* section. In addition many new questions have been added that should help to provide a much stronger connection between the lecture and laboratory portions of the course. Past editions have included the answers to all of the prelaboratory exercises. This edition contains answers to about half of the exercises with the remainder answered in the *Instructor's Manual*. The isolation of the pages of the *Prelaboratory Exercises* resulted in the addition of several blank pages to the book. To fill these pages we have added several provocative and undoubtedly controversial *Chemical Capsules* that hopefully help to demonstrate the importance of chemistry but that also should initiate some productive discussions.

The sequence of experiments and exercises has been altered to provide an even tighter fit with the 5th edition of *Basic Concepts of Chemistry* by Malone and most other preparatory chemistry texts. On the first page of each experiment are listed the topics covered and the corresponding chapters in Malone. Instructors may choose to change the sequence and to select experiments that match their teaching styles and their course content. The experiments can be performed in any order, with the exceptions that *Experiment 5* should follow 4; and 21 and 23 should follow 20 as *Experiments 4* and 20 involve the preparation of materials used in the later experiments.

In addition to sequence changes, there are two notable additions to the *Exercise* section. *Exercise 11* on the periodic properties now includes an optional computer exercise that uses the Project Seraphim program, *K. C. Discover*. *Exercise 29* is a new exercise that reviews and hopefully integrates and reinforces the many concepts covered in the course.

As the first week of most laboratory courses involves check in procedures and safety discussions, we suggest that the first experiment be performed over a two week period. This takes the pressure off of the students during the first day of their laboratory experience and allows the students to perform the experiment at a comfortable and relaxed educational rate.

The *Exercise* section has been included to fill in gaps where experiments are not easily performed and to supplement the problems in the textbooks. Many of these exercises have a multiple choice format so that they can be answered on *Scantron*-type answer sheets, quickly machine corrected by instructors (and recorded as grades if desired) and returned for instant feedback for the students. Most instructors do not have time to grade homework assignments, but this approach provides a possible method. We have included several graphing exercises, as we believe students need more experience with graphing. Also some optional exercise sections have been included in nonmultiple choice format.

Chemistry has been called the "central science" and indeed plays an extremely important role in today's society. Decisions and choices that involve science are made by everyone everyday. Should a person take vitamin C daily or drink diet sodas? Should we continue to burn fossil fuels or should we build more nuclear fission reactors or invest more money in nuclear fusion research? These are questions that rely on some knowledge of science. If our society is to make the wisest decisions and lead us into the 21st century with the best possible lifestyle, we need to have a scientifically literate citizenry. We have attempted to prepare a laboratory text that will provide an environment that will turn you on to science. The joy of discovering on your own and creating experiments to test ideas is truly a rewarding part of life. Begin your experience in the laboratory with a positive attitude and you will find that chemistry is an exciting and enjoyable adventure.

Acknowledgments: We would like to acknowledge the following people for the puns that we have included that they submitted to *Chem 13 News* or *Chemical and Engineering News*: G. White, Avi Ornstein, Mary V. Orna, Ted Jones, Alan Wheatley, Charles Deber, R. J. Morton and John Eix. We would also like to thank the **Milton Roy Company** for their permission to publish the diagram of a Spectronic 20 spectrophotometer. We are grateful to Bonnie Cabot, Clifford Mills and Neda Rose of John Wiley and Sons, Inc., for their assistance in bringing this text to the publication stage. Finally we thank Carolyn, Wendy, Lori, David and Harry (Steve Murov) and Dorothy and John (Brian Stedjee) for their support and encouragement during the preparation of this fourth edition.

Steven Murov
Brian Stedjee

CONTENTS

LABORATORY TECHNIQUES AND SAFETY INFORMATION

1. Wear goggles or appropriate legal eye protection at all times when performing experiments or in the vicinity of experiments being performed.

2. Know the locations and operating instructions for the fire extinguisher, eyewash, safety shower and fire blanket.

3. When heating liquids in test tubes, never point the tube toward yourself or anyone else. Never heat the test tube directly at the bottom but tilt the tube and heat it gently between the bottom of the tube and the top of the liquid.

4. Follow instructions carefully and do not perform unauthorized experiments.

5. Keep table tops clean. Wipe up acid and base spills promptly.

6. Do not take food into the laboratory. Don't taste or eat anything in the laboratory.

7. Fire polish all glass tubing. When inserting glass tubing through a rubber stopper, hold the tubing very close to the rubber stopper with a towel to protect your hand and use glycerin as a lubricant.

8. When working with toxic gases or volatile toxic substances, work in a well operating hood.

9. Report all accidents, no matter how minor, to the instructor.

10. Never return chemicals to bottles of their origin. If you have taken an excess of a chemical, give it to another student, or if necessary, throw the excess away. It's better to waste a small amount of the chemical than to risk contaminating the entire contents of the bottle.

11. Don't stick objects such as pencils or eyedroppers into reagent bottles and don't lay reagent bottle stoppers down in any way that the part which goes into the bottle comes in contact with any surface. If you need a few drops of a liquid, pour a little into a beaker and then take what you need from the beaker. If a solid has packed hard in a bottle, slap the side of the bottle to loosen it. If this doesn't loosen some solid, ask the instructor to help you. Most reagent bottles for solids have hollow caps. If you need a small amount of the solid, with the cap still in the bottle, shake a little into the cap and take what you need from the cap. These techniques prevent introduction of contamination.

12. Switching reagent bottle stoppers will invariably contaminate the reagent. To avoid this, never have more than one bottle unstoppered at a time. If the stopper is the pennyhead type, hold it between the fingers of the hand you are pouring with, while pouring. If you do this, you can be certain you are not mixing up stoppers or contaminating the reagent.

13. Wear a lab apron to protect your skin and clothing whenever you are working with hot or corrosive liquids.

14. Never work alone in the laboratory.

15. Always read the label twice before using a chemical reagent. Be sure the concentration, as well as the name of the reagent, is correct.

16. If you spill a chemical on your skin, flush the area *immediately* with plenty of water, then wash the area with soap and water.

17. Smoking is not permitted in the laboratory.

18. Leather shoes are preferred to canvas or open-toed shoes.

19. Paper items should be disposed of in the waste basket; not in crocks!

20. Do not dispose of insoluble compounds, used metal squares or marble chips in sinks. Dispose of them in designated containers.

Beaker

Clay triangle

Crucible and cover

Evaporating dish

Flame spreader

Erlenmeyer flask

Florence flask

Long-stemmed funnel

Graduated cylinder

Wash bottle

Test tube

Medicine dropper

Watch glass

Crucible tongs

Test tube clamp

Pinch clamp

Wire gauze

4

SAFETY EXERCISE

About 100 years ago, the Chairperson of the Chemistry Department at Johns Hopkins University, and, in large part the founder of chemistry in America, Ira Remsen, wrote the following:[1]

> While reading a textbook of chemistry, I came upon the statement "nitric acid acts upon copper." I was getting tired of reading such absurd stuff and I determined to see what this meant. Copper was more or less familiar to me, for copper cents were then in use. I had seen a bottle marked "nitric acid" on a table in the doctor's office where I was then "doing time!" I did not know its peculiarities, but I was getting on and likely to learn. The spirit of adventure was upon me. Having nitric acid and copper, I had only to learn what the words "act upon" meant. Then the statement, "nitric acid acts upon copper," would be something more than mere words.
>
> All was still. In the interest of knowledge I was even willing to sacrifice one of the few copper cents then in my possession. I put one of them on the table; opened the bottle marked "nitric acid;" poured some of the liquid on the copper; and prepared to make an observation.
>
> But what was this wonderful thing which I beheld? The cent was already changed, and it was no small change either. A greenish-blue liquid foamed and fumed over the cent and over the table. The air in the neighborhood of the performance became dark red. A great colored cloud arose. This was disagreeable and suffocating - how should I stop this? I tried to get rid of the objectionable mess by picking it up and throwing it out of the window, which I had meanwhile opened. I learned another fact - nitric acid not only acts upon copper but it acts upon fingers. The pain led to another unpremeditated experiment. I drew my fingers across my trousers and another fact was discovered. Nitric acid also acts upon trousers. Taking everything into consideration, that was the most impressive experiment, and, relatively, probably the most costly experiment I have ever performed. I tell of it even now with interest. It was a revelation to me. It resulted in a desire on my part to learn more about that remarkable kind of action. Plainly the only way to learn about it was to see its results, to experiment, to work in a laboratory.

The description above is very amusing and expresses an enthusiasm for chemistry that we all should strive for. Ira Remsen also recognized the vital importance of the laboratory experience in chemistry. However, he was very fortunate that this particular experiment did not have dire consequences. Experiments should never be conducted using the methods described. List all the violations of good safety practice in the experiment described by Ira Remsen and suggest some safer approaches to finding out what was meant by the words "acts upon".

---

[1]Getman, F. H., *The Life of Ira Remsen*, Journal of Chem. Ed., Easton, PA, 1940, p 9 (reprinted in Cobb, C., Goldwhite, H., *Creations of Fire: Chemistry's Lively History from Alchemy to the Atomic Age*, Plenum, N.Y. 1995, p. 255).

Experiment 1

INTRODUCTORY CONCEPTS, TECHNIQUES AND CHALLENGES

Learning Objectives

Upon completion of this experiment, students should have learned:
1. The nature of the scientific method.
2. The use of the Bunsen burner.
3. The basic techniques of glassworking.
4. Some differences between metals and nonmetals.
5. The relationship between elements and their compounds.

Text Topics

Scientific method, substances, elements, compounds, mixtures (Malone, *Prologue*, Chapters 2, 3)

Comments

This experiment has been designed for two lab periods. Students should check in on the first laboratory day and after a presentation by the instructor, begin work on the experiment. The experiment can be completed easily on the second lab day. For additional problem solving experience with the nomenclature of elements, see *Exercise 4* on page 305.

Discussion

SCIENCE: THE MEANING AND OPPORTUNITY. Science is a holistic activity. It involves observing, thinking and doing. To learn and understand science, you must not only study it, but also experience it. That is what the laboratory is all about. In this course, you will explore the principles and techniques that enable chemists to probe the unknown, develop theories and make predictions. We urge you to immediately involve yourself by asking about the meaning of science (see *Prelaboratory Exercise* on page 17). Several quotations from recognizable sources concerning science are presented on the next page. Study the quotations, interpret them and try to grasp a feeling for what science is about.

6

"Equipped with his five senses, man explores the universe around him and calls the adventure science." *Edwin Powell Hubble*

"Men love to wonder and that is the seed of our science." *Ralph Waldo Emerson*

"Science is a little bit like the air you breathe - it is everywhere." *Dwight David Eisenhower*

"The most beautiful thing we can experience is the mysterious. It is the source of all true art and science." *Albert Einstein*

"Science is the attempt to make the chaotic diversity of our sense-experience correspond to a logically uniform system of thought." *Albert Einstein*

"Scientific thought is not an accompaniment or condition of human progress, but human progress itself." *William Kingdon Clifford*

"The truth is, that those who have never entered upon scientific pursuits know not a tithe of the poetry by which they are surrounded." *Herbert Spencer*

"Science contributes to our culture in many ways, as a creative intellectual activity in its own right, as the light which has served to illuminate man's place in the universe, and as the source of understanding of man's own nature." *John F. Kennedy*

"People must understand that science is inherently neither a potential for good or for evil. It is a potential to be harnessed by man to do his bidding. Man will determine its direction and its effects. Man, therefore, must understand science if he is to harness it, to live with it, to grow with it." *Glenn T. Seaborg*

"Let both sides seek to invoke the wonders of science instead of its terrors. Together let us explore the stars, conquer the deserts, eradicate disease, tap the ocean depths and encourage the arts and commerce." *John F. Kennedy*

SCIENCE: THE METHOD. An analysis of most scientific discoveries and advances reveals that the process can be broken down into five steps:

1. **observation** - The first step is to notice the universe around us. Puzzling observations lead to questions and this part of the scientific process is the key to good science. The alert observer does not overlook or bypass the puzzles but pauses to determine if a solution to the puzzle should be sought. The process of observation will be expanded upon in the next section.

2. **hypothesis or explanation** - The next step is to develop possible answers or explanations for questions that arise from any unusual, unexplained or unexpected observations.

3. **experiment or test** - One thing that sets science apart from most other fields is that it is usually possible to experimentally test potential answers and explanations.

4. **theory** - A hypothesis is elevated to the level of a theory when experimental testing of a potential answer or explanation supports the hypothesis and enables us to use the theory for predictive purposes.

5. **modification** - Scientific theories are always subject to further testing. When results are obtained that apparently contradict an existing theory, either the theory must be modified and improved or discarded.

It should be emphasized however, that scientists do not stop and ask "which step am I on today?" The five stages of the scientific method are simply a hindsight analysis of the steps that have led to many scientific breakthroughs. For instance, based on observation one might ask why we experience a period of dark and light every 24 hours. Perhaps one might hypothesize that the sun is the source of the light and it goes around the earth every 24 hours. This hypothesis does not become theory as experimental testing causes it to be discarded. Another hypothesis should now be developed that does withstand testing and can be elevated to the level of a theory. The theory should now be useful for prediction.

Observation is the first and key step to the scientific process. As a result of an unusual, unexpected or unexplained observation, we ask questions, propose explanations and then test the potential explanations. Observations must therefore be made carefully and be as complete and unbiased as possible. Observing is more than just seeing, hearing, smelling, feeling or tasting (*PLEASE NOTE RIGHT NOW THAT IN THE CHEMISTRY LABORATORY, WE NEVER TASTE ANYTHING AND SMELLING IS DONE VERY CAREFULLY*). It is being alert for the unusual, unexplained and unexpected so that you will question that which you do not understand. You have certainly observed that ice floats on water but have you ever asked if that is consistent with your expectations (should a solid be more or less dense than the liquid of the same substance)? Observing is noticing and comparing to intuitive expectations. The dictionary[1] defines it this way:

To observe: To see or sense especially through careful analytic attention.

Observation: Act of recognizing and noting some fact or occurrence often involving measurement with instruments.

To gain further insight into the importance of observation, ponder the following quotations:

"Shakespeare says, we are creatures that look before and after: the more surprising that we do not look round a little, and see what is passing under our very eyes." *Thomas Carlyle*

"You see but you do not observe." (Sherlock Holmes to Dr. Watson) *A. Conan Doyle*

"Observation, not old age, brings wisdom." *Publilius Syrus*

"Armado: How has thou purchased this experience?
Moth: By my penny of observation." *William Shakespeare*

"In the fields of observation, chance favors only the mind that is prepared." *Louis Pasteur*

"God hides things by putting them near us." *Ralph Waldo Emerson*

---

[1]Webster's <u>New Collegiate Dictionary</u>, G & C Merriam Co., 1973.

8

In scientific endeavors, we must learn to focus our senses on the experiment and not overlook anything. Often observations that may seem to be irrelevant turn out to be of great importance. Even in our daily lives careful and complete observation can be invaluable. If there is a fire in the laboratory, do you know where the fire extinguishers, fire blankets, and safety showers are? If you don't know, the time lost in finding them might cause someone permanent physical damage or even death. Do you know where alternative exits to the main exit are?

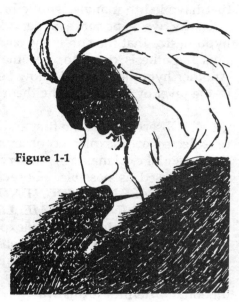

Figure 1-1

Consider the figure to the right. What would you write down in your notebook if you were asked to describe your observations about the figure? If you would have written simply "old woman" or "young woman", it is very possible than you would have been wrong. Look carefully and you will see that it is possible to see both depending on the perspective. The descriptions "old woman" or "young woman" when given alone are conclusions, not observations. Conclusions are also a very important part of a scientific report but they should be labeled as conclusions and kept separate from the pure observations. It would have been all right to have written that the figure appears to be an old woman looking towards the front left and a young woman looking towards the back left but even these descriptions are incomplete observations. The feather and the apparent fur also need to be described. The descriptions that you leave out may be the most important of the possible observations.

When you perform the experiments in this book, try to look for the unusual. Many of the greatest discoveries have been made because someone took the time to investigate small differences between expected and realized results, and other discoveries have been made by scientists who took the extra time to investigate unusual observations. Although the extremely accurate astronomical measurements of Tycho Brahe seemed to confirm the circular planetary orbit theory of Copernicus, Johannes Kepler decided that an eight-minute discrepancy of arc was too much to overlook. As a result, early in the seventeenth century he developed an elliptical model that even today survives as a very accurate mathematical representation of the planetary orbits. Sweeteners such as saccharin and aspartame were both discovered accidentally when researchers noticed a sweet taste on their fingers (bad lab technique) and decided to determine the chemical structure responsible for the taste. Rubber and teflon are two more examples of serendipitous discoveries that were not overlooked by their discoveries but were pursued to a fruitful conclusion.

In addition to being complete, observations should be unbiased. There is a natural tendency to steer an experiment toward expected results. This must be strongly resisted. You should report your observations as accurately and completely as possible without adding your personal interpretation. Some exercises have been designed to illustrate the points just presented and to help you develop and appreciate the value of careful, complete and unbiased observational skills. These exercises are presented in later sections of this experiment.

CHEMISTRY: THE CENTRAL SCIENCE. The branch of science called chemistry has often been described as the central science (why?). The dictionary defines chemistry this way:

A science that deals with the composition, structure and properties of substances and of the transformations that they undergo.

Chemistry is the study of matter and energy from a molecular perspective. This places chemistry in a crucial interdisciplinary position between physics and biology. Consider the number of ways chemistry touches your life everyday. There are man-made chemicals such as cleaning agents, medicines, computer chips, synthetic fabrics, plastic materials, food additives and energy sources. There are entire chemical fields such as pharmacology, nuclear chemistry, biochemistry, genetic engineering, geochemistry and cosmochemistry. We are engulfed in a world of chemistry. In addition and just as important, chemistry is an expression of two of the most important human characteristics; the ability to be curious and the desire to be creative. The experiments in this book have been designed to help you explore and experience the principles and techniques of this very central science.

SUBSTANCES AND MIXTURES: PHYSICAL AND CHEMICAL PROPERTIES. Notice that a key word in the definition of chemistry is the word substance. To the average person, the word substance probably is synonymous with stuff. However, to a chemist, substance has a much more specific meaning; a single chemical. The use of the word substance then implies that the chemical is pure but often to avoid confusion chemists will talk about pure substances. If more than one substance is present, a mixture results. There are two types of substances, elements and compounds. They exist in three common states of matter; solid, liquid or gas depending on the conditions. Compounds are composed of elements chemically combined in definite ratios. One question you will be confronted with in this experiment is the relationship, if any, between the properties of elements and their compounds (do the properties of sodium chloride resemble those of sodium and chlorine?).

Every substance has a unique set of physical and chemical properties that enable us to distinguish, identify and utilize the substance. Physical properties are those that are measured without a change in chemical composition such as melting point, boiling point, color, density and solubility. Chemical properties are those that are determined when a change in chemical composition or structure occurs by decomposition or reaction with other reagents.

Substances usually occur in nature together with other substances as mixtures. Even stockroom chemicals contain a small percentage of impurities ranging from 10% to 0.01%. Mixtures usually must be separated into pure substances before attempts at identification can be initiated. Fortunately mixtures can be separated using procedures that utilize differences in physical properties such as boiling point (distillation) and solubility (recrystallization, filtration and chromatography) to accomplish the separation. A discussion of mixtures and the use of these procedures will follow in later experiments but today's experiment will focus on the qualitative observation of physical and chemical properties of several elements and compounds. Later experiments will focus on the quantitative determination of physical and chemical properties (density, melting point, reaction equilibria, energetics and rates).

One of the ultimate goals of chemistry is to search for order in matter. Are there any similarities among the elements and are there ways of classifying matter to simplify its study. The periodic table is an extremely useful way of arranging the elements and provides us with significant information and predictive ability. Part of this experiment will provide a brief introduction to the periodic table. Refer to the table on the inside front cover of this text for this discussion. Notice the double line staircase that continues with a horizontal segment between boron (B) and aluminum (Al). This line roughly segregates the metals from the nonmetals although some of the elements adjacent to the line are called metalloids and have properties intermediate between metals and nonmetals. The list below compares and contrasts some of the physical properties of metals and nonmetals.

Property	Metal	Nonmetal
appearance	lustrous	nonlustrous
malleability (hammerable)	high	low
ductility (drawable)	high	low
electrical conductance	high	low (insulators)
thermal conductance	high	low
ionization energy	relatively low	relatively high
electron affinity	relatively low	relatively high
electronegativity	relatively low	relatively high
state of matter (room temp.)	solid (except Hg)	10 gases, 1 liquid, 10 solids
reactivity with each other	low	common

Properties of compounds cover a broad range but we do observe that virtually all ionic compounds (metals bonded to nonmetals or polyatomic ions) are solids at room temperature. On the other hand, we find many molecular (or covalently bonded) compounds that exist in the gas or liquid state at room temperature and atmospheric pressure. The goal here will be simply to compare the properties of compounds and their component elements.

In addition to physical properties, you will be observing the chemical properties of some elements and compounds. Chemical reactions and properties are usually detected when one of the following is observed: formation of an insoluble product (precipitate), evolution of gas (bubbles) or change of temperature or color.

Procedure

A. THE BURNER AND GLASSWORKING. The instructor should demonstrate the use of a Bunsen burner and its controls. To show the hottest regions of the flame, an index card can be inserted carefully and perpendicularly into the flame (slightly tilted). Additionally it is possible to insert an unused match into the flame directly above the burner head and show that the stick will ignite near the exterior of the flame before the head which is in a relatively cool part of the flame. The instructor should also demonstrate the glassworking techniques of firepolishing and glass bending. After the demonstration, students should make the pieces of glassware in *Figure 1-2* as selected by the instructor.

The burner. There are two adjustments on the Bunsen burner (see *Figure 1-3*). The needle valve on the bottom adjusts the gas flow. The air vents that are adjusted by rotating the barrel of the burner regulate the amount of air that is mixed with the gas before it is burned. When the air vents are entirely closed, the burner produces a yellow, smoky flame known as a luminous flame. If the air vents are open too far, the flame will go out. When the burner is adjusted properly, the flame will be nonluminous and totally blue. There should be an inner blue cone about 2 to 3 cm high and an outer purple envelope. There should be a mild hiss and no yellow color. With the air vents nearly closed, light the burner. Then adjust the burner to give the appropriate flame.

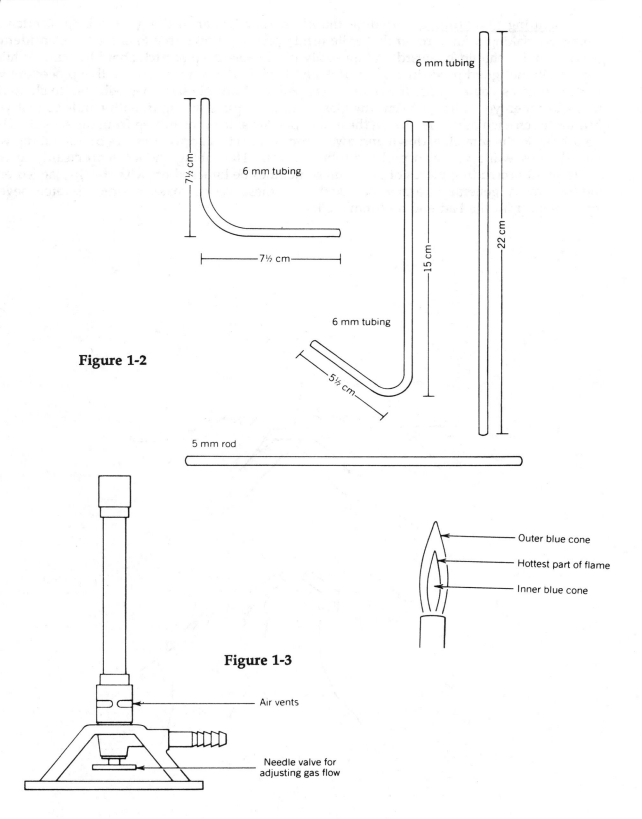

Figure 1-2

Figure 1-3

12

Cutting glass tubing. Holding the file perpendicular to the glass tubing, scratch the tubing by making a forward stroke while firmly pressing down (see *Figure 1-4*). Considerable grinding noise should be heard and an easily observable deep scratch should result. While it is generally not good procedure to go back and forth to deepen the scratch, this procedure will work as long as you *keep the file in the same groove*. Multiple scratches will lead to shattering or cracks when you attempt to cut the glass. Hold the glass tubing in both hands so that your thumbs touch each other and are on the exact opposite side of the tubing from the scratch. Hold the tubing so the scratch is down and away from you. Then apply pressure on the tubing with both thumbs while you also pull the tubing apart. The tubing should snap cleanly in two. While small protruding pieces of glass can sometimes be knocked off with the file, jagged ends and/or cracks generally cannot be fixed. In these cases, make another scratch several centimeters from the bad end and remove it.

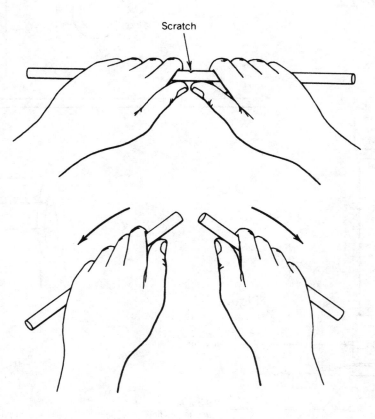

Figure 1-4

Fire polishing. Glass tubing that has just been cut has sharp edges that make it difficult to insert through rubber stoppers and easy to cut fingers. For these reasons the ends of the cut glass tubing are always fire polished. Fire polishing is accomplished by holding the end of the tubing about 30° to the horizontal in the hottest part of the flame while slowly rotating it (see *Figure 1-5*). The flame above the glass tubing should take on a yellow color. After a short time, the edges of the end of the tube should appear less angular. Don't heat the tubing too long or the opening will begin to constrict. When you finish firepolishing a piece of tubing, lay the hot end on your wire gauze pad to protect it from thermal shock and to protect the table top.

Bending glass tubing. To bend glass tubing properly, you must put your flame spreader on the burner. Turn off the burner before you do this. Relight the burner and adjust it so the flame appears as a blue fringe. Holding the tubing at both ends and lengthwise with respect to the flame, lower it to the top of the blue region of the flame and rotate it slowly (see *Figure 1-6*). A uniform yellow flame above the tubing indicates it is being heated evenly. When the tubing appears to have softened, remove it from the flame and bend it. A proper bend is smooth rather than sharp. Lay the hot part of the glass tubing on your wire gauze pad to cool.

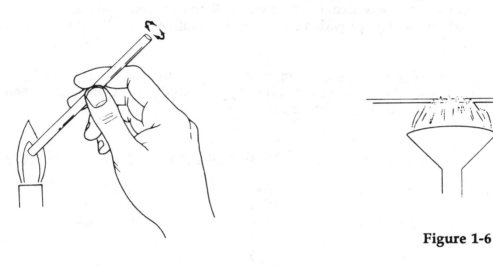

Figure 1-6

Figure 1-5

B. OBSERVATIONS AND THE SCIENTIFIC METHOD.

1. The meniscus. Add water to about the half way point of a 50 mL graduated cylinder and study the features of the water surface. Describe and draw your observations. The phenomenon that you observe is called a meniscus. Write down a question about the meniscus. *[Comment: Volumetric glassware has been calibrated to give correct volume measurements when you read the very bottom of the meniscus.]*

2. Stirring device. You will make a very brief application of the scientific method to the stirring device (the electrical, mechanical device that the instructor has set up, not a stirring rod.) that you will find in the lab. Fill out the accompanying question sheet (page 19) and proceed through an application of the scientific method. This involves observing and questioning, suggesting an explanation for any puzzling observations and testing your explanations.

14

C. PHYSICAL PROPERTIES OF SUBSTANCES.

1. Solids, liquids and gases. Place a few pieces of ice in a 400 mL beaker. In a second 400 mL beaker, add about 200 mL of water. Compare some easily observable properties of the solid and liquid states of water (relationship of shape to shape of container, does it flow, etc.). Now place the beaker of water on a wire gauze supported above a Bunsen burner by a ring attached to a ring stand. Heat the water to boiling and observe the boiling process. What are the bubbles and what shape would the gaseous water assume if trapped in a container? How much volume would the bubbles occupy if all the liquid were to be converted to gas? Consider the difference in the relative positions and the movement of the molecules in each phase of water. Do not turn off the flame but continue right on to the next part with the boiling water.

2. Molecular motion in liquids. To one 150 mL beaker *(using beaker tongs)*, add about 100 mL of the boiling water from C-1 above. To a second, identical 150 mL beaker add 100 mL of water straight from the tap. Carefully add one drop of food coloring in the same manner to each beaker and report and explain your observations.

3. Metals, nonmetals and metalloid. About ten test tubes labeled with code letters and melting point, density and electrical conductivity will be available in a test tube rack for your inspection. Based on the data provided and your observations, classify the elements as metal, nonmetal or metalloid.

4. Elements and compounds. Samples of the substances listed in the table below will be available in sealed vials. Using the data supplied and visual inspection, compare the compounds to the elements that compose them.

Substance	Density (g/cm^3)	Melting Point (°C)	Water Solubility (g/100 g soln.)
iron	7.86	1536	insoluble
iron(II) sulfide	4.74	1193	6.2×10^{-4}
sulfur	2.07	119	insoluble
zinc	7.14	419	insoluble
zinc iodide	4.74	446	83
iodine	4.94	114	0.029

5. Compounds and mixtures. A compound consists of elements combined chemically in a specific ratio. Compounds can be broken down by chemical means and the result is either new compounds or elements. In sharp contrast to compounds, mixtures have variable compositions. Different amounts of salt can be dissolved in water to form salt water solutions with different concentrations of salt in the solution. Unlike compounds, mixtures can be separated into substances using physical means. In this procedure, you will make some simple observations on solutions of copper sulfate and on the dependence of the solubility of this compound on temperature.

Add about 200 mL of water to a 400 mL beaker. Place the beaker on a hot plate or on a wire gauze supported above a Bunsen burner and heat the water to boiling. *[Caution: Always use beaker tongs when manipulating beakers of hot liquids. Do not use crucible tongs for beakers.]* While waiting for the water to boil, add about 0.1 gram of copper sulfate pentahydrate ($CuSO_4 \cdot 5H_2O$) to about 5 mL of water in a 18×150 mm test tube. Mix the contents of the tube by firmly grasping the test tube between your thumb and forefinger of one hand and striking the bottom of the test tube vigorously and frequently with the forefinger of your other hand. *Never* put a thumb over the mouth of the test tube to avoid spilling when shaking. If the method above does not achieve adequate mixing, insert a cork or rubber stopper into the tube and then shake as you hold the stopper down with your thumb. After the $CuSO_4 \cdot 5H_2O$ dissolves, observe the color and its intensity. Add an additional 1 gram of $CuSO_4 \cdot 5H_2O$ and repeat the dissolving and observing process. Now add an additional 2.5 grams of $CuSO_4 \cdot 5H_2O$ to the solution and attempt to repeat the dissolving process. If it doesn't dissolve, put the test tube in the beaker of boiling water and stir the mixture in the test tube with a stirring rod until the copper sulfate pentahydrate dissolves. Place the test tube in a rack or beaker and allow it to cool to room temperature. If nothing happens, scratch the inside of the tube with a glass rod. Report your observations.

D.   CHEMICAL REACTIONS.   Upon the mixing of two solutions, the four common observations that indicate that a chemical change has occurred are: formation of an insoluble product (precipitate), bubbles (evolution of a gas), temperature change or a color change. The absence of all of these observations often but not always means that there has not been a chemical reaction. Negative results are just as important as positive results and must be appropriately recorded. In fact, the absence of a reaction is a very common observation. You will prepare 5 different mixtures, make observations and determine if a reaction has occurred. Pour about 2 mL of the first solution into a test tube and add about 2 mL of the second solution, mix and observe.

System	Solution 1	Solution 2
A	calcium chloride (0.1 M)	sodium carbonate (0.1 M)
B	hydrochloric acid (3 M)	sodium hydroxide (3 M)
C	calcium chloride (0.1 M)	potassium nitrate (0.1 M)
D	sodium carbonate (1 M)	hydrochloric acid (3 M)
E	iron(III) chloride (0.1 M)	potassium thiocyanate (0.1 M)

E.   MYSTERY FLASK.   Add the solutions below to a 250 mL Erlenmeyer flask, swirl until mixed and allow to stand without agitation until an observable change occurs (several minutes).

15 mL of a dextrose (glucose) solution (80 g/L)
15 mL of a potassium hydroxide solution (64 g/L)
10 drops of a methylene blue solution (0.4 g/L)

Record your observations. Be sure to look at the flask from many angles. Vigorously swirl the solution for several seconds and again record your observations. Repeat the sequence as often as you desire but focus your attention on the change that occurs when you swirl the mixture.

(chemical pun) *What does ABCDEFGHIJKLMNOPQSTUVWXYZ represent?   [Hint: it is the third most abundant gas in the atmosphere.]   For answer, see next page.*

Chemical Capsule

Argon (the r is missing or gone) is the third most abundant gas (not counting water which varies) in the atmosphere (volume percents; nitrogen - 78.09%, oxygen - 20.95%, argon - 0.93%, carbon dioxide - 0.03%). The name argon is derived from the Greek, argos which appropriately means lazy or inactive. Argon along with helium (Greek - sun), neon (Greek - new), krypton (Greek - hidden), xenon (Greek - stranger) and radon (ray) make up the group or family of elements called inert or noble gases. As Mendeleev's first version of the periodic table was developed in 1869, only a few months after as the discovery of the first inert gas (helium), inert gases were not included in the first periodic table. In fact, most of the other inert gases were not discovered until 1898 by Sir William Ramsay.

In 1962 Neil Bartlett demonstrated that inert gases are not completely inert when he synthesized $XePtF_6$. Shortly thereafter, chemists were able to make several fluorides of xenon and krypton.

Because of their extremely low chemical reactivity, the inert gases have many important practical uses. Neon is used in "neon" lighting and argon is used to provide an inert atmosphere in electric light bulbs and for chemical reactions. The wavelength of an orange-red line in the atomic spectrum of krypton is used to define the meter. Helium, because of its low density is used for balloons and blimps. It is sometimes mixed with oxygen as a breathing mixture for divers. Perhaps even more important, helium's very low boiling point of 4.2 K makes liquid helium a commonly used coolant when very low temperatures are needed (e.g., to provide a temperature where some materials become superconducting).

On a lighter side, krypton is probably most commonly recognized as the name of Superman's home planet. The substance, "kryptonite", is said to be a green solid that is potentially lethal to Superman but harmless to humans. In *Superman III*, the computer scientist (portrayed by Richard Pryor) analyzed "kryptonite" and found it to consist of 15.08% plutonium, 18.06% tantalum, 27.71% xenon, 24.02% promethium, 10.62% dialium, 3.94% mercury and 0.57% unknown. Would this mixture be harmless to humans?

Prelaboratory Exercises - *Experiment 1*

For solutions to the starred problems, see *Appendix A*.

Name_____Date_____Lab Section_____

1. Read the quotations on science and make your own quotable comment about science.

2.* Your lab drawer contains beakers and Erlenmeyer flasks of similar sizes. Why are you supplied two different shaped containers for the same volume? Give advantages and disadvantages of each type of container.

3. What color is water?   _____

4.* Name the elements you can think of that are found in elemental form (uncombined with other elements) on earth.

    _____ _____ _____ _____ _____ _____ _____

    _____ _____ _____ _____ _____ _____ _____

5. Classify the following as elements, compounds or mixtures.

    a.*   aluminum   _____

    b.*   water (pure)   _____

    c.*   air   _____

    d.   carbon dioxide   _____

    e.   milk   _____

    f.   gold (24 carat)   _____

    g.   glucose   _____

    h.   vinegar   _____

6. Consider the properties of the compound, sodium chloride, and the elements sodium and chlorine. Does sodium chloride have properties in common with sodium and/or chlorine?

7. Are the following changes physical or chemical?

    a.* The boiling of water   _____

    b.* The reaction of iron with oxygen to form rust (Fe_2O_3)   _____

    c.* The evaporation of rubbing alcohol   _____

    d.   Frying an egg   _____

    e.   Burning of a candle   _____

    f.   The melting of ice   _____

8.* Notice that the dictionary definition of chemistry included the two words composition and structure. Structures of three compounds with the formula (composition) $C_2H_2Cl_2$ are given below along with some properties of the two compounds. What is the significance of the word structure in the definition of chemistry?

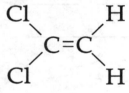
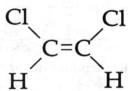

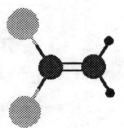
1,1-dichloroethene

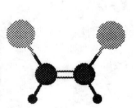
cis-1,2-dichloroethene

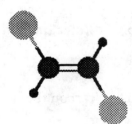
trans-1,2-dichloroethene.

	1,1-dichloroethene	cis-1,2-dichloroethene	trans-1,2-dichloroethene
b.p.	37°C	60°C	48°C
m.p.	-122°C	-80°C	-50°C
density	1.218 g/mL	1.284 g/mL	1.257 g/mL

18

Results and Discussion - *Experiment 1*
INTRODUCTORY CONCEPTS, TECHNIQUES AND CHALLENGES

Name_____Date_____Lab Section_____

A. THE BURNER AND GLASSWORKING.

1. In the diagram, label:
   a. the gas control
   b. the air control
   c. the hottest part of the flame

2. Show the results of your glassworking to your instructor

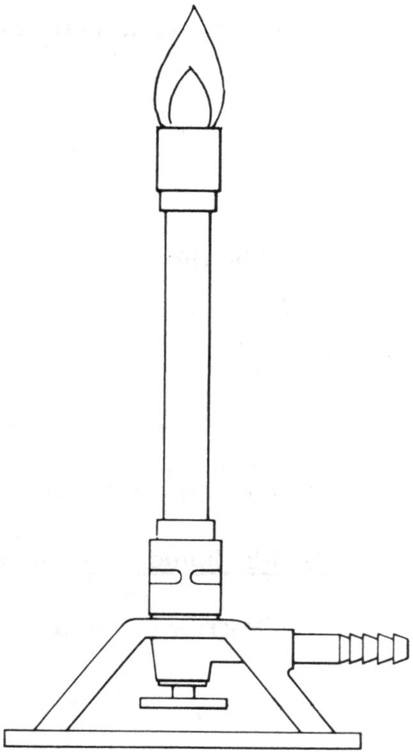

Figure 1-7

B. OBSERVATIONS AND THE SCIENTIFIC METHOD.

1. <u>The meniscus.</u> Describe and draw the meniscus.

2. <u>Stirring device.</u>

   a. Write down all observations of the device you can make without manipulating it.

   b. What question should be asked about the device?

c. Suggest an explanation to answer the question in #2-*b* above.

d. Describe and carry out an experiment to test your hypothesis in #2-*c* above.

e. Give the results of #2-*d* above. Was your explanation consistent with the results?

C. PHYSICAL PROPERTIES OF SUBSTANCES.

1. Solids, liquids and gases.

a. Describe some of the differences between the solid, liquid and gaseous states.

b. What are the bubbles in the boiling water? From the volume of the bubbles, can you comment on the relative densities of liquid water and gaseous water.

c. (Bonus question) Consider the term "liquid crystal". Where are they encountered and why are they called liquid crystals?

2. Molecular motion in liquids.

a. Compare and contrast your observations about the behavior of food coloring in hot and room temperature water.

b. Explain your observations in *2-a* above.

c. If a salt solution (at room temperature) is poured into water (at room temperature), is it necessary to stir to quickly achieve thorough mixing? Explain your answer.

3. Metals, nonmetals and metalloid. Classify each tube as metal (M), nonmetal (NM) or metalloid (MD).

Tube	Classification	Tube	Classification
A		F	
B		G	
C		H	
D		I	
E		J	

4. Elements and compounds.

a. Does there appear to be any correlation between the properties of a compound and its elements? Explain your answer.

b. Shortly after the Chernobyl nuclear reactor accident, some people took potassium iodide tablets to dilute the radioactive iodide in their bodies and diminish its retention in the thyroid gland. Although many experts questioned this practice, little harm was probably caused because of the relatively low toxicity of KI. However, on May 12, 1986, *Newsweek* incorrectly captioned a photo that showed a child apparently receiving KI with "On Alert: Administering iodine to Polish children." Critically evaluate the mistake and possible consequences.

5. Compounds and mixtures.

a. Describe the test tube contents after you have added 0.1 gram of copper sulfate pentahydrate and stirred.

b. Describe any significant changes after an additional 1 gram of $CuSO_4 \cdot 5H_2O$ is added and mixed.

c. How can the copper(II) sulfate - water system be distinguished from a compound?

d. Describe your observations when an additional 2.5 grams of $CuSO_4 \cdot 5H_2O$ is added to the solution.

e. What happens to the mixture in *5-d* when it is heated?

f. Describe what happens when the solution from *5-e* is allowed to cool. Are your observations consistent with your expectations? Explain your answer.

D. CHEMICAL REACTIONS.

<u>Reaction</u>   <u>Observations</u>

A

B

C

D

E

E. MYSTERY FLASK.

1. Observations upon standing:

2. Observations upon swirling:

3. Suggest an explanation for the change that occurs upon swirling.

4. How could you test your explanation?

Suggest any ways you can think of to improve any part(s) of this experiment.

Experiment 2

MEASUREMENTS

Learning Objectives

Upon completion of this experiment, students should have learned:
1. To use and read thermometers, balances, graduated cylinders, and pipets.
2. To apply significant figures.
3. The meanings of "precision" and "accuracy".

Text Topics

Metric system, significant figures, unit conversions, temperature, accuracy and precision (Malone, Chapter 1)

Comments

For additional problem solving experiences involving significant figures, scientific, notation and unit conversions, see *Exercises 1, 2*.

Discussion

In all sciences, measurements are essential. The most fundamental properties that can be measured are length, mass, and time. In chemistry, temperature is also often treated as a fundamental property. Other properties of matter such as volume, density or velocity are ratios or products of the fundamental properties. For instance:

1. Units of volume are length^3
2. Units for density are mass/length^3.
3. Units for velocity are length/time.

The metric system is used almost exclusively in all sciences. Historically the metric system developed in two forms. Internationally, the SI (System International) has been adopted as the standard measurement system. The meter, kilogram and the second are the basic units in this system but the meter and the kilogram are generally too large for convenient use in a chemistry laboratory. The form that is usually used in chemistry is known as the CGS system where the basic units of length, mass and time are the centimeter, gram and second respectively.

Units in the metric system are related to each other as multiples of 10 and designated by prefixes. Some of the most commonly used prefixes are:

pico	-	trillionth	10^{-12}
nano	-	**billionth**	10^{-9}
micro	-	**millionth**	10^{-6}
milli	-	**thousandth**	10^{-3}
centi	-	**hundredth**	10^{-2}
deci	-	tenth	10^{-1}
kilo	-	**thousand**	10^{3}
mega	-	million	10^{6}

These prefixes are attached to the terms which differentiate between the different types of measurement: meter for length, gram for mass, second for time, liter for volume, etc. Thus, a centimeter is a hundreth of a meter, a kilogram is a thousand grams, a milliliter is a thousandth of a liter.

This experiment has been designed to acquaint you with several types of measurement and measuring devices. The first lesson you must learn when using any measuring device is to always estimate and record data to one digit beyond the smallest graduation on the measuring device. Your estimate has significance and to ignore it is to overlook a useful digit that may be important in the analysis of results. Readings on state of the art digital instrumentation such as electronic balances sometimes provide the exception to the rule that you estimate one digit beyond the last graduation.

In the length measurement below it is easy to see that the marker is between 1.3 and 1.4 units. It appears to be closer to 1.3 units than 1.4 units. However, to write down simply 1.3 units is really stating that the value is between 1.25 and 1.35 units when by more careful observation, you can see that it is between 1.31 and 1.34 units. Imagine that the interval 1.3 to 1.4 is divided into tenths and you will probably conclude the best estimate for the value is 1.32 or 1.33 units.

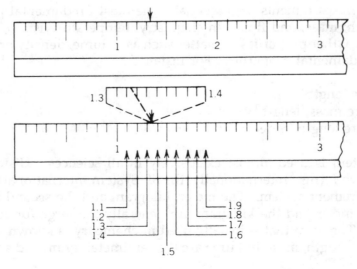

Figure 2-1

When reading a volume of water in a container such as a graduated cylinder, you will notice (as in *Experiment 1*) that the water is higher at the edges than in the middle and forms a phenomenon called a meniscus. Always read the low point of the meniscus. Confirm in your mind that the two values below are 24.0 mL and 16.6 mL, respectively. Notice the zero for 24.0 *must* be recorded. If you record only 24 mL, you are saying the value is between 23.5 and 24.5 mL; whereas, 24.0 mL means between 23.95 and 24.05 mL. Notice also, recording of the units is just as important as recording the number. Do you agree that the thermometer in *Figure 2-3* indicates 13.3°C?

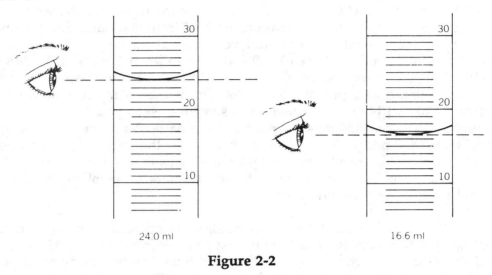

24.0 ml   16.6 ml

Figure 2-2

Figure 2-3

SCIENTIFIC NOTATION. In the sciences it is very common to encounter very small and very large numbers (e.g., the diameter of an atom is about 0.0000000001 m and 12 grams of carbon-12 contains 602,213,700,000,000,000,000,000 atoms). Because of the inconvenience of writing such numbers and to avoid the confusion of significant figures in numbers such as 320, scientists usually express numbers in scientific notation. To write a number in scientific notation:

1. Move the decimal point so that it becomes a number between one and ten:

    63,000 becomes 6.3, 0.0035 becomes 3.5

2. If, when the decimal point is moved, the resulting number, between one and ten, is smaller than the original number (6.3 is smaller than 63,000) the exponent will be the number of places the decimal was moved.

    63,000 is written 6.3×10^4 in scientific notation.
    1 2 3 4

    If, when the decimal point is moved, the resulting number, between one and ten, is larger than the original number (3.5 is larger than 0.0035), the exponent will be minus the number of places the decimal was moved.

    0.0035 is written 3.5×10^{-3} in scientific notation.
    1 2 3

3. Whenever a decimal is moved, if the resulting number is larger than the original, the exponent is made smaller by the same number of places.

$$0.012 \times 10^{-5} = 1.2 \times 10^{-7}$$

Whenever a decimal is moved and the resulting number is smaller than the original, the exponent is made larger by the same number of places.

SIGNIFICANT FIGURES. It is necessary, in science, to convey the accuracy of figures reported in data and the accuracy of conclusions derived from that data. Suppose as you drive your car the speedometer reads 65 mph and you drive for 2 hours. Could you say with confidence that you had driven between 129.9 and 130.1 miles or 129 and 131 miles or 120 and 140 miles. Because of several reasons (there are substantial errors in speedometer readings, possible errors in timing and probable deviations from the 65 mph), you would probably respond between 120 and 140 miles. However, if someone asked you how far you had driven, you would probably respond 130 miles no matter which particular range was most appropriate. For scientists, this is not adequate as scientists must know the error range of a measurement. When constructing a table, would it be all right for the legs to differ in length by 1 mm, 1 cm or 1 m? Probably a 1 mm tolerance would not lead to significant wobbling but legs that differ by 1 cm would make a very wobbly table.

The question then is how do you communicate the precision of the number to other people. Perhaps the best way is to include the range of possible values (130 ± 10 miles) but determining standard deviations takes some time and often is more information than is really needed. To provide a very simple method that admittedly does have some shortcomings, scientists have developed a convention based on *significant figures.* Use of this method enables the users to instantly ascertain an idea of the range of possible values of a measurement or a value calculated from a series of measurements. For the distance problem in the paragraph above, a range between 129.9 and 130.1 would be expressed 130.0 miles or 1.300×10^2 miles, between 129 and 131 would be expressed 1.30×10^2 miles (as you will learn, 130 miles leads to ambiguity and it is necessary to use scientific notation to avoid confusion over significant figures when there is no decimal showing and the last digit is a zero), and between 120 to 140 miles would be expressed 1.3×10^2 miles (although 130 miles would be interpreted the same way).

In a measurement such as the 1.32 units on page 26, there are three significant figures. The 1 and 3 were read directly from graduations on the measuring instrument and the 2 was estimated but provides the useful information that the actual value probably lies in the range 1.31 to 1.33 units. To be able to take advantage of the significant figure method of keeping track of precision, it is necessary to learn the set of conventions that have been established for their use.

All recorded digits read directly, plus the final estimated digit are significant. The primary difficulty with significant figures is with zeros. Zeros have two roles; numerical value and decimal placement. When used as numerical values, they are significant. Zeros are not significant when used as decimal place holders. Consider a measurement of 1.2 cm. This can also be expressed as 0.012 m. Clearly, changing units cannot change the number of significant figures. As 1.2 cm has two significant figures, 0.012 cm must also have two. The first zero is present for aesthetic reasons and the second simply to place the decimal. In a number such as 10.010, there are five significant figures as all have numerical value with the zero on the far right an estimated digit.

Numbers such as 430 that end in zero and do not have a decimal showing are the major source of confusion when determining significant figures. Here the zero could be an estimated digit or simply filling a decimal place. To avoid this confusion, this number should be written 4.3×10^2 when the 3 is the estimated digit and 4.30×10^2 when the 0 is the estimated digit. The guidelines summarize methods for determining if zeros are significant.

1.   If no decimal is present, count the number of digits from right to left starting and ending with non-zero digits

2.   If a decimal is present, count all digits left to right starting with the first non-zero digit. In this case, count all zeros to the right of the first digit.

	Significant Figures
5027	4
0.0129	3
0.320	3
23.0	3
5270	3

Rules for significant figures during mathematical operations. When multiplying or dividing, the answer should have the same number of significant figures as the piece of information with the fewest number of significant figures.

$$(1.50 \text{ grams NaOH}) \left(\frac{1 \text{ mole NaOH}}{39.9971 \text{ g NaOH}} \right) = 0.0375027 \text{ moles NaOH} = 0.0375 \text{ moles NaOH}$$

(1.50 grams NaOH) — 3 sig. figs.
(39.9971 g NaOH) — 6 sig. figs.
0.0375027 moles NaOH — (too many sig. figs.)
0.0375 moles NaOH — (3 sig. figs.)

When adding or subtracting, the answer can be no more precise than the least precise piece of information.

Formula mass of PbO
lead	207.19	g/mol
oxygen	15.9994	g/mol

| PbO | 223.1894 g/mol | = **223.19 g/mol** |

As the digits beyond the hundredths place are not known for lead, the value for lead oxide cannot be given with any confidence beyond the hundreth place.

DIMENSIONAL ANALYSIS (also known as unit conversions, factor label method). Problems in chemistry often involve many steps. Application of a technique called dimensional analysis can often provide a systematic simplified approach towards the solution of these problems. In dimensional analysis, a measured or given value is multiplied by a unit conversion or series of unit conversions until a new numerical value is obtained with the desired units. Consider for example, the equality, 100 cm = 1 m. This equation can be divided on both sides by 100 cm and the equation 1 = 1 m/100 cm results. The ratio 1 m/100 cm is equal to 1 and is therefore called a unit conversion. As any measurement can be multiplied by the number 1 without changing the true value of the measurement, multiplication by unit conversions simply changes the measurement from one set of units to another. Notice also that the reciprocal of 1 is still 1 so any unit conversion can be inverted and used as a unit conversion (100 cm/1 m is a unit conversion).

For example, if you are 1.70×10^2 cm tall and want to know the value in meters, multiply by the unit conversion 1 m/100 cm.

$$(1.70 \times 10^2 \text{ cm})\left(\frac{1 \text{ m}}{100 \text{ cm}}\right) = 1.70 \text{ m}$$

Notice that the goal is to use the unit conversion that will cancel the given units (cm) and introduce the desired units (m). When the unit conversion or ratio of desired units to given units is not immediately available, it is sometimes possible to string a series of unit conversions together to obtain the desired units. Now lets check the *saying* that "a pint is a pound the world around" by utilizing the density of water at 20°C (0.998 g/mL) in the metric system. Notice in the following calculation that sequential diagonal cancellation of units using familiar unit conversions is utilized until the desired result is obtained.

$$(1.00 \text{ pt})\left(\frac{1 \text{ qt}}{2 \text{ pt}}\right)\left(\frac{1 \text{ L}}{1.057 \text{ qt}}\right)\left(\frac{10^3 \text{ mL}}{1 \text{ L}}\right)\left(\frac{0.998 \text{ g}}{1 \text{ mL}}\right)\left(\frac{1 \text{ lb}}{453.7 \text{ g}}\right) = 1.04 \text{ lb}$$

The *saying* apparently has a 4% error.

The sequence that you use to perform the above calculation is not relevant as long as values in the numerator are multiplied and values in the denominator are used to divide. Be sure to write the complete unit. If your problem is concerned with grams of copper, the unit is written "grams Cu", not simply "grams". "Grams Cu" does not cancel with "grams Zn".

Alternate procedure for dimensional analysis. Some people find that it is easier to work backward from the unit or units of the answer than to work from the data given to towards the answer.

Question: How many millimeters are there in 2.6 miles?

Step 1:	Write one unit of the answer on both sides of the equal sign.	mm = mm
Step 2:	In the setup, fill in the information or conversion factor that contains the unit of the answer.	$\frac{10 \text{ mm}}{1 \text{ cm}}$ = mm
Step 3:	Write the unit of the setup that doesn't occur in the answer in the place where it will cancel:	$\text{cm} \times \frac{10 \text{ mm}}{1 \text{ cm}}$ = mm
Step 4:	In the setup, fill in the information or conversion factor that contains that unit.	$\frac{2.54 \text{ cm}}{1 \text{ in}} \times \frac{10 \text{ mm}}{1 \text{ cm}}$ = mm
Step 5:	Continue this process until all units cancel except those in the answer.	

$$2.6 \text{ miles} \times \frac{5280 \text{ ft}}{1 \text{ mile}} \times \frac{12 \text{ in}}{1 \text{ ft}} \times \frac{2.54 \text{ cm}}{1 \text{ in}} \times \frac{10 \text{ mm}}{1 \text{ cm}} = \text{mm}$$

Procedure

PART A - TEMPERATURE MEASUREMENTS AND EFFECTS

1. Using your thermometer, determine the temperature of some tap water. Be sure of three points:

    a. The mercury is not moving when you make the measurement.
    b. The mercury bulb is totally in the water and not touching the glass.
    c. The reading is estimated to one digit beyond the last graduation (probably 0.1°C)

2. Bring about 200 mL of water in a 400 mL beaker to a boil and measure the temperature of the boiling water.

3. a. Make about 20 mL of an ice slush (about 25% water, 75% ice) in a small beaker (about 50 or 100 mL). After stirring, measure the temperature of the slush. Save the slush for *Part A-3b* (next).

    b. Add about 5 grams of sodium chloride to the slush and stir. Measure the temperature of the slush.

PART B - MASS MEASUREMENTS

Make sure that you are familiar with the instructions for the balances that are available to you.

1. Three objects; a rubber stopper, a ball bearing and a cork ring will be available in a container near the balances. Before weighing the objects, pick up each object and attempt to estimate the weight relative to the other objects. Then weigh each object to 0.01 grams or 0.001 grams depending on the limits of your balance.

2. Weigh a small beaker. Then add 20 drops of water to it with your medicine dropper and weigh the beaker and its contents. The purpose of this experiment is to find the approximate number of drops in a milliliter and the volume of a drop of water.

PART C - VOLUME, ACCURACY AND PRECISION

Scientific experiments are not entirely free from errors. Errors can result from limitations of an experimenter's equipment or abilities. It is an extremely rare occurrence when an experimental result turns out to be exactly the result that was predicted theoretically. The measure of how close an experimental result is to a theoretically predicted result is called *"accuracy"*. Some kinds of equipment are designed to be more accurate than others. For example, one can measure 10 milliliters of a liquid using either a 10 mL volumetric pipet or graduated cylinder but the pipet will nearly always be more accurate.

Because errors are unavoidable a scientist always runs an experiment several times and averages the results. This average is normally more accurate than the individual results. If an experiment is a good one, the results the experimenter obtains should all be close to one another. The measure of how closely experimental results agree is called *"precision"*. The number of significant figures present is generally a measure of precision. If one measuring device can be read to 25.0 mL and a second to 25.00 mL, the latter should be able to deliver amounts that are closer together (more precise) each time. Another method of evaluating precision is to determine the average deviation. First calculate the average of several determinations and sum the differences between each measurement and the average value. Divide the sum of the deviations by the number of determinations to get the average deviation. See *Problem 6* in the *Prelaboratory Exercises* for an example of the technique.

Weigh a dry 150 mL beaker to at least the nearest centigram. Now measure 10 mL of deionized water in your graduated cylinder. Notice the meniscus and be sure to read the very bottom of the meniscus and avoid parallax error. Add the 10 mL of water to the beaker and weigh the beaker accurately again. Use the graduated cylinder again to measure 10 mL of water, add the water to the beaker, and weigh again. Repeat this process a third time.

Using a pipet bulb, draw deionized water (not from the beaker you just weighed!) into a clean 10 mL volumetric pipet until it is a few centimeters above the line (see *Figure 2-4* on the next page). Quickly remove the bulb and place your index finger firmly over the opening. As you decrease your finger pressure on the opening of the pipet, air will enter it and the liquid level will drop. Allow it to drop until the bottom of the meniscus is exactly on the calibration line. Then reapply finger pressure so that the liquid level doesn't drop any further. Now add the water from the pipet to the beaker containing water from the graduated cylinder. When most but not all of the water has drained from the pipet, touch the pipet tip onto the inside wall of the beaker (above the water level) and allow the pipet to continue to drain naturally (do not blow the last drop out). After draining is complete, hold the tip in contact with the inside wall of the beaker for another 10 seconds. If you have pipetted properly, there should be a small drop of water remaining in the pipet tip but no water hanging from the tip. The pipet was calibrated assuming that the small drop would be left in the tip. Weigh the beaker accurately. Pipet an additional 10 mL of water and weigh again. Repeat the process a third time.

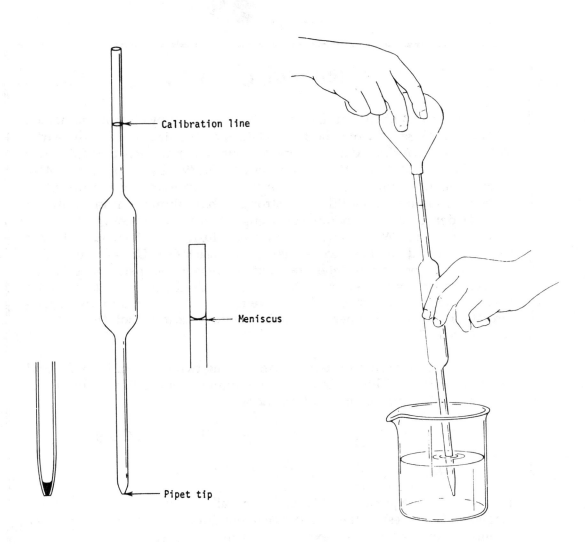

Calibration line

Meniscus

Pipet tip

Figure 2-4

Chemical Capsule

The element, titanium, was discovered in 1791 by Gregor and named in 1795 by Klaproth from the Latin, Titans or the first sons of the earth. This could be considered an appropriate name as titanium is the ninth most abundant element in the earths crust (0.63% by weight). More impressively, it ranks second to iron in abundance among the transition metals. Titanium is significantly stronger than aluminum and only 1.7 times as dense. It is comparable in strength to iron but has only 60% of the density of iron. When pure, titanium is a lustrous, white metal that is easily fabricated and has excellent corrosion resistance. Unlike many other metals whose properties deteriorate with increasing temperature, titanium retains most of its properties up to several hundred degrees Celsius. With these wonderful properties, why do we find its use limited to high speed aircraft, rockets and other applications where cost is not the primary consideration?

Titanium occurs in nature primarily as two ores, rutile (TiO_2) and ilmenite ($FeTiO_3$). The elemental form of titanium is produced from rutile using the following sequence of reactions:

$$TiO_2(s) \; + \; 2\,C(s) \; + \; 2\,Cl_2(g) \; = \; TiCl_4(g) \; + \; 2\,CO(g)$$

$$TiCl_4(g) \; + \; 2\,Mg(s) \; = \; Ti(s) \; + \; 2\,MgCl_2(s)$$

This reaction sequence and alternatives all involve use of expensive reagents. Thus, despite the abundance of titanium and its useful properties, titanium remains today too expensive for most applications.

On the other hand, because of the abundance and properties of rutile, TiO_2 is used frequently. TiO_2 is brilliantly white, very stable, opaque and very importantly, has low toxicity. These properties make it an ideal pigmentation for paints and the annual production in the United States is near one million tons. Unfortunately, many paints used to contain lead compounds. Some children still get lead poisoning from eating pealed off paint chips.

Prelaboratory Exercises - *Experiment 2*

For solutions to the starred problems, see *Appendix A*. For additional problems on significant figures, scientific notation and unit conversions, see *Exercises 1, 2*.

Name_____Date_____Lab Section_____

1. a.* Give four advantages of the metric system over the "American" system of measurement.

    b. Give the pros and cons of dividing an hour into 100 minutes instead of 60 minutes.

    c. Give the pros and cons of dividing a circle into 400° instead of 360°.

2. How many significant figures are in each of the following:

    a.* 305 cm   _____   d. 0.0125 kg   _____   g. 3.00×10^{-1} L _____

    b.* 0.70 g   _____   e. 9760 m   _____   h. 20.50 mL   _____

    c.* 4.600×10^3 L _____   f. 2×10^3 mg _____   i. 0.05050 g   _____

3. Perform the following conversions or calculations. Use unit conversions when appropriate, show your work and use the correct number of significant figures.

    a.* 5.2 g $\rightarrow$ kg   _____   g.* (4.75 cm)(2.0 cm) =   _____

    b.* 3.74×10^{-1} km $\rightarrow$ cm _____   h.* $\dfrac{1.640\times10^3 \text{ g}}{8.13\times10^1 \text{ mol}}$ =   _____

    c.* 7.5×10^1 cm^2 $\rightarrow$ m^2 _____   i.* 9876 cm + 1.2 cm =   _____

    d. 6.21×10^{-2} L $\rightarrow$ mL   _____   j. 1.5×10^3 m + 2.1×10^2m = _____

    e. 4.0×10^2 mg $\rightarrow$ kg   _____   k. $\dfrac{(4.20\times10^{-2})(2.7\times10^{-1})}{(9.0\times10^1)(7.0\times10^{-6})}$ = _____

    f. 1.4 mL $\rightarrow$ m^3 _____   l. $(12.011)/(6.022\times10^{23})$ =   _____

4. a. Measure and record the length and width (to the perforations) in cm of this piece of paper.

length _____

width _____

   b. Calculate the area of this piece of paper in m². _____

   c. Give instructions for determining the thickness of this piece of paper in cm. Perform the measurements and calculations and report your answer. _____

5.* The density of ethanol is 0.80 g/mL.

   a. If 15 drops of ethanol from a medicine dropper weigh 0.60 grams, how many drops does it take from the dropper to dispense 1.0 mL of ethanol?

   b. What is the volume in milliliters of 1 drop of ethanol?

6.* The mass of a 3.500 gram piece of iron is measured three times on two different balances with the following results:

Trial	Balance 1 (grams)	Balance 2 (grams)
1	3.53	3.37
2	3.55	3.45
3	3.51	3.47

   a. Calculate the average for each set of measurements on each balance.

   _____(1)

   _____(2)

   b. Calculate the average deviation for each set of measurements on each balance. _____(1)

   _____(2)

   c. Which balance is more precise? Explain your answer

   d. Which balance is more accurate? Explain your answer

36

Results and Discussion - *Experiment 2*
MEASUREMENTS

Name_____Date_____Lab Section_____

A. TEMPERATURE MEASUREMENTS AND EFFECTS

1. Water at room temperature _____

2. Boiling water _____

3. Ice water

    a. After stirring _____

    b. With salt added _____

    c. The reason for observation *3b* is _____

        1. Salt freezes at a lower temperature than water

        2. Salt water freezes at a lower temperature than water

        3. Friction caused the ice to melt

        4. Salt raises the freezing point of water

        5. None of the above

B. MASS MEASUREMENTS

For the first three items, estimate relative masses by holding the items in your hand before you actually measure the masses.

container number   _____

	Estimated Ranking (1-3, 1 = heaviest)	Measured Mass (grams)	Actual Ranking (1-3, 1 = heaviest)
1. Rubber stopper	_____	_____	_____
2. Ball bearing	_____	_____	_____
3. Cork	_____	_____	_____

4. Why do the estimated rankings differ from the actual rankings?   _____

    a. The hand measures weight, not mass
    b. The hand approximately measures density (mass/volume)
    c. The hand approximately measures pressure (mass/area)
    d. The hand measures squeezability

Dropper experiment

5. Mass of the container   _____

6. Mass of container + 20 drops of water   _____

7. Mass of 20 drops of water   _____

8. Using the results from #7 and the density of water (1.0 g/mL), calculate (do not measure) the number of drops in 1 mL of water.

show calculation   _____ × _____ = _____

9. Using the results from #7 and the density of water, calculate the volume in milliliters of one drop of water.

show calculation   _____ × _____ = _____

10. Which measuring device in your lab drawer would you use to measure approximately ¼ mL of a liquid?   _____

C. VOLUME, ACCURACY AND PRECISION

1.

		graduated cylinder (g)	pipet (g)
a.	Mass of beaker and contents before addition of water	_____	_____
b.	After first addition	_____	_____
c.	After second addition	_____	_____
d.	After third addition	_____	_____
e.	Mass of first 10 mL of water	_____	_____
f.	Mass of second 10 mL of water	_____	_____
g.	Mass of third 10 mL of water	_____	_____
h.	Average of 3 trials for graduated cylinder		_____
i.	Average of 3 trials for pipet		_____
j.	Deviation of first trial from average	_____	_____
k.	Deviation of second trial from average	_____	_____
l.	Deviation of third trial from average	_____	_____
m.	Average deviation for graduated cylinder		_____
n.	Average deviation for pipet		_____

2. Did you have better precision using the graduated cylinder or the pipet? Explain your answer. _____

3. Using appropriate table in the *Handbook of Chemistry and Physics*, report the density of water at the temperature of water measured in # *A-1* (edition_____, page_____)   _____

4. Using the density of water from # *C-3* above, calculate the mass of 10.00 mL of water at the temperature of # *A-1*. Show your calculations.   _____

5. Calculate the percentage deviation between the theoretical value (*C-4*) and each of your measured values (*C-1h and C-1i*) for the graduated cylinder and the pipet (show your calculations).

    % deviation for the graduated cylinder   _____

    % deviation for the pipet   _____

6. Which measuring device was more accurate according to your results. Explain your answer.   _____

Experiment 3

DENSITY

Learning Objectives

Upon completion of this experiment, students should have learned:
1. Concepts of density, mass and volume.
2. To determine the densities of liquids and solids.
3. Fundamentals of graphing.
4. To apply the scientific method.

Text Topics

Density, graphing (Malone, Chapter 2, *Appendix D*)

Comments

For additional problem solving experience with density, see *Exercises 2, 3.*

Discussion

The density of a substance is defined as the mass divided by the volume: $d=m/v$. Density is a physical property of a substance that does not depend on the amount of material present and is therefore called an intensive property. 1.00 mL of mercury has a mass of 13.6 grams and a density of 13.6 grams/1.00 mL = 13.6 g/mL. 10.0 mL of mercury has a mass of 136 grams and a density of 136 grams/10.0 mL = 13.6 g/mL. Because a pure substance at a given pressure and temperature has a fixed density, the density often assists in the identification of a substance. A measured density of 13.6 g/mL provides strong supporting evidence that the substance is mercury. Density is also a parameter that enables you to calculate the mass of a given volume of substance or the volume of a given mass of the substance. As density = mass/volume, the determination of density involves the measurement of both the mass and volume of a substance.

You should try to obtain a sense for typical values of densities of common substances. Examine the following table:

substance	symbol	state (1 atm., 20°C)	density (g/cm³)
gold	Au	solid	19.3
lead	Pb	solid	11.3
copper	Cu	solid	8.92
copper sulfate	$CuSO_4$	solid	3.6
copper sulfate pentahydrate	$CuSO_4 \cdot 5H_2O$	solid	2.28
iron	Fe	solid	7.86
iron(III) oxide (rust)	Fe_2O_3	solid	5.24
carbon (diamond)	C	solid	3.51
carbon (graphite)	C	solid	2.25
aluminum	Al	solid	2.70
sodium chloride	NaCl	solid	2.17
sucrose (table sugar)	$C_{12}H_{22}O_{11}$	solid	1.58
water	H_2O	liquid	0.998
ethanol	C_2H_6O	liquid	0.789
octane	C_8H_{18}	liquid	0.703
oxygen	O_2	gas	0.00132
nitrogen	N_2	gas	0.00116
helium	He	gas	0.000166

Note that liquids and solids are hundreds of times denser than gases.

Density determinations are also performed on mixtures of two substances to determine the percent composition. The density depends on the percent composition of the mixture and varies between the densities of the two pure substances. Density measurements of pure substances and mixtures of known composition can be made and a table or graph prepared. The density of a mixture of unknown composition can then be measured and its percentage composition can be determined from the table or graph. A density measurement on your car's coolant can be used to determine the percent of antifreeze (ethylene glycol) in the coolant. The density of your car battery solution enables a mechanic to determine if the battery is still good.

GRAPHING. It is very common for scientific experiments to result in tables full of data. While it is sometimes possible to come to meaningful conclusions from tables, it is often easier to discern relationships between variables visually using appropriate graphs. A graph shows how one variable changes as another is varied. A graph should be designed to be easily read and interpreted. The guidelines below should help you prepare readable graphs.

Graphing hints

1. Select logical scales that utilize as much of the graph paper as possible. For many graphs, the two axes will have different scales. Each major division should represent 1, 2, 2.5 or 5 units or some power of ten times one of these units. Never use 3, 7 or another number that makes it difficult to locate points on the graph.

2. Label the axes and indicate units used.

3. Locate the points and put a dot at the proper locations. *Circle* the dots.

4. If the points appear to fall on a straight line, use a *straight edge* to draw the line that best fits (averages the deviations) the data. Do not sketch straight lines and do not connect the dots with straight line segments. If the points seem to fall on a smooth curve draw the best curve possible through the data.

5. Title the graph.

Graphs are often a good method of testing scientific theories. Suppose, for example, that you derive the relationship that the distance you travel in a car should equal the velocity of the car multiplied by the time, $d = vt$. To test the equation, drive a car at a constant velocity and measure the time at various distances.

distance (miles)	time (hours)	distance/time (miles/hr)
13.5	0.25	54.0
34.2	0.60	57.0
56.0	1.00	56.0
74.2	1.40	53.0
99.0	1.80	55.0

The equation can be checked by dividing the distance by the time (d/t) and comparing the values as in the table above. As you can observe, the d/t values (or the velocity) came out very close to a constant value. This strongly supports the validity of the equation $d = vt$. An alternative method of checking the equation is to graph d versus t as shown below. If $d = vt$ is true, a straight line should result. The general equation for a straight line is $y = mx + b$ where y is the dependent variable, x the independent variable, m the slope (rise/run) of the line, and b the intercept. For the equation $d = vt$, $d = y$, $v = m$, $t = x$ and $b = 0$. The graph shows that within experimental error a straight line does result with a slope equal to the average velocity and a zero intercept. This again supports the validity of the equation $d = vt$.

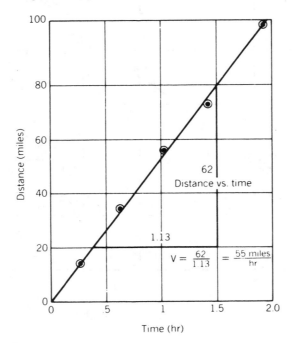

Figure 3-1

Distance vs. time

$V = \frac{62}{1.13} = \frac{55 \text{ miles}}{hr}$

Suppose you want to find out if the volume of a gas (at fixed pressure) is related by an equation of the type y = mx + b to the temperature of the gas (V = mt + b ?). You measure the volume at several temperatures and obtain the following data:

Temperature (°C)	Volume (Liters)
100	2.96
80	2.87
60	2.68
40	2.53
20	2.33
0	2.21
-20	2.07
-40	1.88

To graph the data, you must first select appropriate scales for the axes. The vertical scale will represent the volume and varies from 1.88 to 2.96 liters. The vertical scale on our piece of graph paper has 7 major divisions. Dividing the volume range of 1.08 liters (2.96 - 1.88) by 7 divisions gives 0.154 liters per division. Using 0.154 liters/division would make it next to impossible to locate the position of the points for plotting and also very difficult for others to read. Instead, the next largest convenient ratio is used, 0.2 liters/division. It should be noted that the volume does not have to be graphed from 0.00 - 2.96 liters as we are interested in this case in finding out if the data fits the equation V = mt + b for the volume range 1.88 to 2.96 liters.

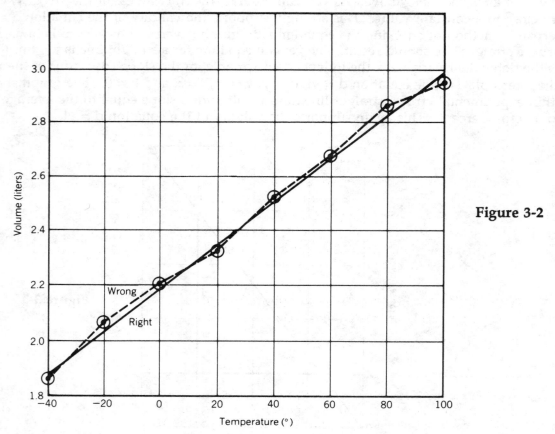

Figure 3-2

The horizontal scale represents the temperature. Seven divisions for a 140°C range gives 20°C per division. The points are plotted (see *Figure 3-2*) and circled and the best straight line through the data is drawn using a straight edge if the data appear to fall on a straight line. Do not connect the dots. The scatter is due to experimental error. The points fall very close to a straight line and it is safe to assume that $V = mt + b$. We will pursue this relationship further in *Exercise 19*.

In many systems, two variables are related to each other by an equation that is not in the form $y = mx + b$. In these cases, a plot of y versus x does not give a straight line and it is difficult to graphically test the validity of an equation. It is sometimes possible to rearrange nonlinear relationships into the form $y = mx + b$ so that a check for a straight line graph can be made. For example, the volume of a gas is inversely proportional to the pressure ($V = m/P$). A plot of V versus P gives a hyperbola rather than straight line and it is difficult to determine that the curve is indeed a hyperbola. However, by letting $x = 1/P$, the equation becomes $V = mx$ which is in the form $y = mx + b$ (with $b = 0$). The equation can then be tested by plotting V versus x ($1/P$).

Graphical plots that give curves also can be used for informative purposes. In the event that the points appear to fall on a smooth curve, draw the best curve you can attempting to average out the experimental deviations from the curve. As a very interesting relevant example, consider the carbon dioxide content of the atmosphere (in ppm - parts per million) versus year given in the table[1].

year	CO_2 conc. (ppm)	year	CO_2 conc. (ppm)
1740	276.9	1940	308.1
1780	279.2	1950	311.0
1820	283.5	1960	316.5
1860	288.4	1970	324.7
1880	291.9	1975	330.2
1900	295.4	1980	337.6
1910	298.5	1985	344.7
1920	302.0	1988	350.0
1930	305.3	1991	354.2

We want the vertical axis to represent the carbon dioxide concentration and vary from at least 270 to 360 ppm or a 90 ppm range. Assuming the graph paper has nine major vertical divisions, we arrive at 90/9 or 10 ppm/division. The horizontal axis will represent the year and should span at least 260 years. The graph paper at our disposal has twelve major horizontal divisions. Instead of using 260/12 or about 23 years/division, for ease and convenience, 25 years/division is selected.

What significant observations can you make from the carbon dioxide graph? You may want to do some reading on the Greenhouse effect and its possible consequences on the global environment.

---

[1]We would like to thank Tim Whorf and C. D. Keeling of the Geological Research Division of Scripps Institution of Oceanography for providing this data. The data up to 1955 represent data obtained from polar ice core samples. The data from 1955 on are from combined atmospheric measurements made at Mauna Loa Observatory in Hawaii and at the South Pole.

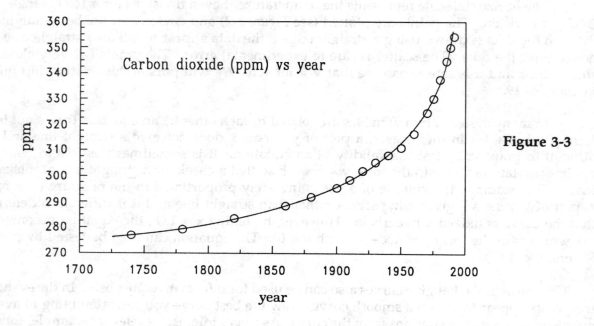

Figure 3-3

Procedure

A. DENSITY OF WATER. In the first part of this experiment, you will determine the densities of three different volumes of water.

1. Weigh an empty, *dry* 50 mL graduated cylinder .

2. Add exactly 10.0 mL of water to the cylinder. Remember, the *bottom* of the meniscus should just be touching the 10.0 mL line. [*Hint:* Add water up to about the 9 mL mark and use a dropper to reach the 10.0 mL mark.]

3. Weigh the cylinder + 10 mL of water. You can now calculate the density of the water by determining the mass of the water (mass of cylinder + water - mass of cylinder) and dividing it by 10.0 mL.

4. Add water up to the 30.0 mL mark and weigh.

5. Add water up to the 50.0 mL mark and weigh.

6. Calculate the densities of the three volumes.

7. As a beginning exercise in graphing, plot the masses of the water on the vertical scale versus the volumes of the water on the horizontal scale.

B. DENSITY OF A SOLID OBJECT. Use your balance to determine the mass of a metal cylinder. The volume of the cylinder will be determined using two different methods. The density (d = m/V) will be calculated and compared for the two volume measurements.

1. Volume from linear measurements:

Perhaps the simplest method of finding the volume of an object involves linear measurement. For instance, the volume of a rectangular solid can be found by multiplying its height times its length times its width. The volume of a cylinder can be found by multiplying π times the square of the radius times the height, $V = \pi r^2 H$. The radius is ½ the diameter and π is approximately equal to 3.14. Calculate the volume using the method above that applies.

2. Volume from water displacement:

The method above is only useful for regular solids whose volumes are related by a formula to the measurements. The volume of an irregular solid must be determined using a different method. One such method utilizes the displacement of a liquid. Half fill your graduated cylinder with water. Read the volume of water to 0.1 mL Slant your cylinder and slowly slide your solid object into it. The object should be totally immersed. Again, read the level of the liquid to the nearest 0.1 mL. The difference in the levels of the liquid equals the volume of the solid object.

C. DENSITY OF A SALT SOLUTION. The density versus percent concentration of several aqueous sodium chloride solutions are provided in the **Results and Discussion** section. Prepare a graph by plotting the density on the vertical (y axis) and the percent by mass of sodium chloride on the horizontal axis (x axis).

Weigh a 150 mL beaker to at least the nearest 0.01 gram and pipet 10.00 mL of your unknown sodium chloride solution into the beaker. Weigh the beaker and contents and calculate the density of the salt solution. Use the graph to determine the mass percent of sodium chloride in the salt solution. [Note: The densities presented of the salt solutions were measured at 20°C. Although densities are temperature dependent, the values are accurate to three significant figures after the decimal between 18 and 24°C. If your solution temperature is not in this range, there will be a slight error.]

D. QUALITATIVE OBSERVATIONS ON DENSITY: AN APPLICATION OF THE SCIENTIFIC METHOD. There should be four beakers in the laboratory available for your inspection. The first beaker labeled "acetophenone", contains <u>*only*</u> acetophenone in its solid and liquid phases. The second beaker, labeled "water", contains only water in its solid and liquid phases. Compare and contrast the properties of the two substances.

The third and fourth beakers contain only water and a piece of aluminum. The masses of the aluminum in the two beakers are the same. Compare and contrast your observations on the contents of the two beakers.

What is $AlPO_4$?

What is HIJKLMNO?

Chemical Capsule

Most people when asked what the three most abundant gases in the atmosphere are include carbon dioxide among their answers. They believe this because plants use CO_2 and animals exhale it. Actually, the air exhaled by humans is only about 3% CO_2 and plants are able to get enough CO_2 for their needs despite the fact that CO_2 ranks a distant 4th (not counting water) in atmospheric abundance at only 0.03%.

Because the amount of carbon dioxide in the atmosphere is so low, the burning of fossil fuels to produce primarily carbon dioxide and water has been able to significantly increase the amount of carbon dioxide in the atmosphere (by about 25% - see page 45 for additional information). As carbon dioxide contributes to the so called "Greenhouse effect", scientists are now very concerned that our burning of fossil fuels will lead to a warming of the earth by a few degrees and potentially very damaging changes in microclimates. Consider the consequences if this turns out to be true including effects on agriculture. What alternative energy sources could be substituted for fossil fuels in the event we find that this is a real threat to the future of society?

You are probably familiar with the solid state of CO_2, so-called dry ice. At –78.5°C, dry ice undergoes a phase change (called sublimation) directly into the gas phase. To get liquid carbon dioxide, the dry ice must be under a pressure of several atmospheres. Then warming of the dry ice results in melting instead of sublimation.

You should also recognize carbon dioxide as the bubbles in carbonated soda and the chemical in many fire extinguishers. Why is it useful in fire extinguishers?

Dogs

Water - H_2O (H to O)

Prelaboratory Exercises - *Experiment 3*

For solutions to the starred problems, see *Appendix A*. For additional problems involving density, see *Exercises 2, 3*.

Name_____Date_____Lab Section_____

1.* a. To a graduated cylinder (mass = 89.22 g), you add ethanol until
        the 25.0 mL mark is reached. The cylinder and contents now weigh
        108.95 g. What is the density of ethanol?   _____

    b. Use the density from *1-a* as a unit conversion to calculate the
        volume necessary to provide 5.00 g of ethanol.   _____

2. a. To a graduated cylinder (mass = 35.89 g), you add mercury until
        the 5.00 mL mark is reached. The cylinder and contents now weigh
        103.89 g. What is the density of mercury?   _____

    b. Use the density from *2-a* as a unit conversion to calculate the
        mass necessary to provide 1.50 mL of mercury.   _____

3.* A 12.5 cm diameter piece of *Whatman #1* filter paper has a thickness
    of 0.162 mm and a mass of 1.089 g. What is the density of the filter
    paper?   _____

4. A 7.61 cm x 7.68 cm rectangular piece of glass has a thickness of 2.2 mm
    and a mass of 35.303 g. What is the density of the glass?   _____

5. A 24.3 g irregular nugget displaces 1.26 mL of water. Is it likely that the nugget is gold? Explain your answer.

6. Could the density of sugar be determined by the water displacement method? If not, suggest a method that could be used. Explain your answer.

7. Could the density of a cork be determined by the water displacement method? If not, suggest a method that could be used. Explain your answer.

8.* Water and antifreeze (ethylene glycol) mixtures have the densities listed in the table below. Plot the density on the vertical axis and the mass % ethylene glycol on the horizontal axis and determine the % ethylene glycol content of an ethylene glycol-water mixture that has a density of 1.030 g/mL.

% water	% ethylene glycol	density (g/mL)
100	0	0.998
90	10	1.011
80	20	1.024
68	32	1.040
60	40	1.051
52	48	1.067

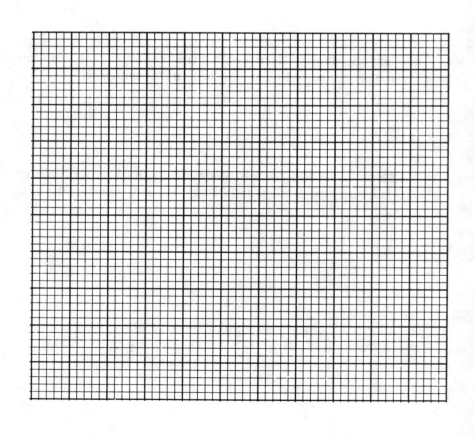

Results and Discussion - *Experiment 3* - DENSITY

Name_____Date_____Lab Section_____

A. DENSITY OF WATER

	10.0 mL	30.0 mL	50.0 mL
1. Mass of graduated cylinder + water	_____	_____	_____
2. Mass of empty graduated cylinder	_____	_____	_____
3. Mass of water	_____	_____	_____
4. Density of 10.0 mL sample of water	_____		
5. Density of 30.0 mL sample of water		_____	
6. Density of 50.0 mL sample of water			_____

7. Show calculation setups below:

10.0 mL

30.0 mL

50.0 mL

Graph mass on the vertical axis vs. volume on the horizontal axis below:

8. Does the density of water change as its mass changes? Explain your answer.

B. DENSITY OF METAL CYLINDER

1. Linear measurement method

   a. Number of metal cylinder _____

   b. Mass of metal cylinder _____

   c. Height of metal cylinder _____

   d. Diameter of metal cylinder _____

   e. Radius of metal cylinder _____

   f. Volume = $\pi r^2 H$ (show calculations) _____

   g. Density of metal cylinder (show calculations) _____

2. Volume measurement method

   a. Mass of metal cylinder _____

   b. Initial volume of water in graduated cylinder _____

   c. Final volume of water in graduated cylinder _____

   d. Volume of metal cylinder (show calculations) _____

   e. Density of metal cylinder (show calculations) _____

3. Suggest reasons for the differences between the two density measurements.

C. DENSITY OF A SALT SOLUTION

1. Unknown number   _____

2. Mass of beaker   _____

3. Mass of beaker + 10.00 mL of unknown   _____

4. Mass of 10.00 mL of unknown   _____

5. Density of unknown   _____

6. The densities of several water - sodium chloride mixtures are reported below. Plot the % by mass of sodium chloride on the horizontal (x) axis and the density on the vertical (y) axis.

% NaCl	density (g/mL)
0.00	0.998
5.00	1.034
10.00	1.071
15.00	1.108
20.00	1.148
25.00	1.189

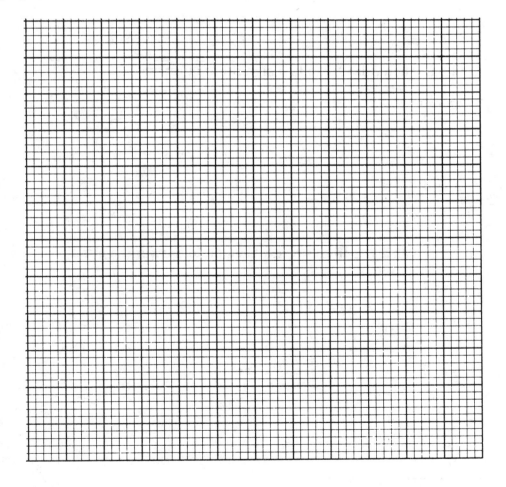

7. From the graph, determine the percent by mass of sodium chloride in your unknown.   _____

D. QUALITATIVE OBSERVATIONS ON DENSITY: AN APPLICATION OF THE SCIENTIFIC METHOD

Beakers containing the following will be available for observation:

1. acetophenone[1]
2. water
3. water + aluminum[2]
4. water + aluminum[2]

Observe the four beakers and compare and contrast *1* with *2* and *3* with *4*.

a. Observations
   1 vs 2

   3 vs 4

Each pair should lead you to a significant question.

b. Questions:
   1 vs 2

   3 vs 4

c. Formulate a hypothesis or explanation to answer the questions.

   1 vs 2

   3 vs 4

---

[1]The only substance (excluding air) in beaker #1 is acetophenone. In laboratories with inadequate heating it may be necessary to replace acetophenone with acetic acid.

[2]The pieces of aluminum in both beakers have the same mass.

Experiment 4

MIXTURES: SEPARATION AND PURIFICATION

Learning Objectives

Upon completion of this experiment, students should have learned:
1. The difference between a substance and a mixture.
2. The terminology of solutions.
3. To separate mixtures by using gravity and vacuum filtrations.
4. To purify a solid using recrystallization.
5. To determine the concentration of a solution by weighing, evaporating to dryness and reweighing.
6. The general effect of temperature on the solubility of organic compounds.

Text Topics

Homogeneous and heterogeneous mixtures, utilization of physical properties to achieve separation and purification (Malone, Chapter 3)

Comments

The percent recovery and purity of the vanillin purified in today's experiment will be determined in next week's experiment so the vanillin purified today must be saved for next week.

Discussion

Recall from the *Discussion* in *Experiment 1* that chemists use the word substance to refer to a pure chemical. There are two kinds of substances: elements and compounds. When two or more substances are physically combined, a mixture results. If sugar is dissolved in water, a homogeneous mixture or solution is produced. You should be aware that most solutions you deal with are predominantly water. In contrast to a sugar - water solution, orange juice contains discernible particles. Anytime it can be demonstrated that a mixture is not the same throughout, it is said to be a heterogeneous mixture. If a microscope reveals that different substances are present or if light is scattered as it passes through a liquid mixture, the mixture is heterogeneous. A mixture of salt and sand may appear to be the same throughout but it is heterogeneous.

Freshly opened carbonated drinks are heterogenous and homogenized milk is heterogenous as it will scatter light. The dairy industry uses the word homogenized in a different way than chemists.

One of the most important tasks in chemistry is the separation of mixtures into pure substances. It is usually possible to separate mixtures into substances by taking advantage of differences in physical properties. Since the boiling point of water (100°C) is much lower than that of sodium chloride (1413°C), you will be able to use evaporation to isolate sodium chloride from its aqueous solution. Knowing the mass of the solution before evaporation and the mass of the residue, you will be able to calculate the original concentration of a sodium chloride solution. In today's experiment, the sodium chloride solution will be one that is saturated; it contains the maximum mass percentage of sodium chloride that will dissolve in water at room temperature.

$$\frac{\text{grams of sodium chloride residue}}{\text{grams of solution}} \times 100\% = \text{mass \% NaCl}$$

In the second part of this experiment, you will mix homogeneous aqueous solutions of sodium carbonate and calcium chloride. The positive ions (sodium and calcium) will switch partners with the negative ions (carbonate and chloride) resulting in the formation of sodium chloride and calcium carbonate.

$$Na_2CO_3(aq) + CaCl_2(aq) = 2\,NaCl(aq) + CaCO_3(s)$$

A heterogeneous mixture will be produced as the calcium carbonate (an ingredient in a number of commercial products such as paint, antacids and chalk) is insoluble in water and precipitates out from the solution. It is possible to separate the calcium carbonate from the aqueous solution by using gravity filtration. The process above can be used as a synthesis and isolation of calcium carbonate. When a synthesis has been completed, it is important to verify the identity of the product. One technique used to characterize carbonates is to add acid. Some kinds of compounds do effervesce or fizz when acid is added, but if no effervescence occurs, it is evidence that the compound is not a carbonate. The potential reaction is given below.

$$CaCO_3(s) + 2\,HCl(aq) = CaCl_2(aq) + H_2O(l) + CO_2(g)$$

The third part of this experiment involves the use of the technique called recrystallization. This technique utilizes the observation that the solubility of solids in liquids usually increases dramatically as the solvent temperature is increased. A barely saturated solution of the solid in the hot solvent is prepared and allowed to cool. The solubility decreases as the temperature drops. Solid crystallizes out of the solution and is filtered and dried. Insoluble impurities are removed by filtration of the hot saturated solution and soluble impurities dissolve in the solvent and do not crystallize back out upon cooling. These impurities pass through the final cold filtration. Several steps are followed in a typical recrystallization.

A. Solvent selection - The solubility of the solid is tested in possible solvents until desirable solvent properties are found (solid should have low solubility in the cold solvent and high solubility in the hot solvent).

B. Dissolving of solid - The solid is dissolved in a minimum amount of hot (usually boiling) solvent to prepare a saturated solution.

C. Hot filtration - If insoluble impurities are present (not in today's experiment) a hot gravity filtration is performed to collect the impurities.

D. Solution cooling - The solution is allowed to cool and solidification should occur. Sometimes seeding, scratching or cooling in an ice bath is necessary to induce crystallization.

E. Ice bath cooling - The solution is usually cooled with an ice bath to further decrease the solubility of the solid in the solvent.

F. Vacuum filtration - A Buchner funnel, a filter flask, and an aspirator are used to collect the solid.

G. Washing the crystals - While the crystals are still in the Buchner funnel, they are usually washed with ice-cold solvent.

H. Drying - The crystals are usually air dried to allow evaporation of the solvent.

I. Yield determination - The mass of the recrystallized sample divided by the mass of the starting solid times 100% is the % recovery.

J. Purity determination - The purity of the sample is usually determined by a simple physical measurement such as melting point.

We began this discussion by commenting on the importance of separation and purification techniques. Most samples encountered by chemists are mixtures and must be separated before analysis can be attempted. Even stockroom chemicals should be checked for purity before use. Stockroom chemicals are purchased from chemical companies who either extract them from natural sources or synthesize them (prepare them using a chemical reaction) from other compounds. Syntheses and/or extractions seldom yield pure products. As a consequence, commercially available chemicals come in different grades of purity and are often only 90% - 95% pure. For some applications, 95% may be satisfactory while other applications require further purification. This experiment will introduce you to the separation and purification techniques of filtration, evaporation and recrystallization. Future experiments will deal with two other common purification techniques, chromatography and distillation. For purification of solids, however, the method you will use in this experiment, recrystallization, is routinely used and generally the first method attempted.

Procedure

A. EVAPORATION.

The amount of sodium chloride dissolved in a saturated solution can be determined by evaporating a weighed amount of a saturated solution to dryness. Dryness is confirmed by repeated heating and weighing cycles until a constant mass is achieved (usually only two cycles are necessary.

1. Weigh a clean, dry evaporating dish to at least the nearest 0.01 gram.

2. Carefully pour about 6 mL of a saturated sodium chloride solution into a graduated cylinder. Do not stir the solution before pouring. Leave the solid NaCl as undisturbed as possible in the bottom of its bottle. Transfer the 6 mL of NaCl solution to the evaporating dish and weigh the dish and its contents.

3. Put a 400 mL beaker about half full of water on a wire gauze above a Bunsen burner. Boil the water and suspend the evaporating dish in the 400 mL beaker. Boil the water (adding water as needed to the beaker to maintain a reasonable level) until the evaporating dish attains apparent dryness (about 20 minutes).

4. Using beaker tongs, remove the beaker and evaporating dish from the flame. Holding it with crucible tongs, place the evaporating dish on a wire gauze and gently flame it for about 3 minutes and allow it to cool.

5. Weigh the dish to at least the nearest 0.01 gram. Again flame the evaporating dish gently (about 3 minutes), cool and weigh. Repeat the process until successive weighing differences are less than 0.02 gram. Calculate the mass percent of NaCl in the saturated solution.

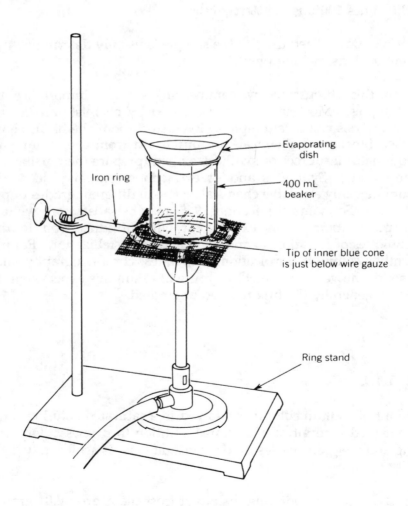

Figure 4-1

B. FILTRATION

1. Fold a piece of filter paper in half. Tear off about a half centimeter piece from one corner and then fold it into quarters. Open up one pocket of the filter paper so that it forms the shape of a cone. Put it into your funnel and wet it thoroughly with deionized water from your wash bottle so that it adheres uniformly to the inside wall of the funnel.

2. Put 10 mL of 1.0 M sodium carbonate into your 150 mL beaker. Add 10 mL of 1.0 M calcium chloride. Record your observations.

3. Swirl the resulting mixture for a few seconds and filter it. When all the aqueous phase has drained, open up the filter paper and scrape the precipitate onto a watch glass with a stirring rod.

4. Using a medicine dropper, rapidly add about 10 drops of 6 M HCl to the precipitate. Describe your observations.

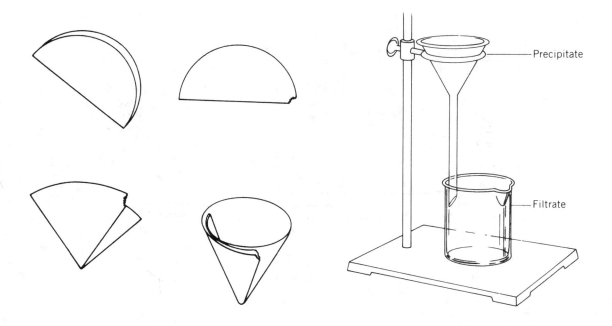

Figure 4-2

C. RECRYSTALLIZATION OF VANILLIN (4-hydroxy-3-methoxybenzaldehyde)

Assume you work for a flavor company and a shipment of vanillin comes in that you suspect is impure. Your task this week is to purify the vanillin by *recrystallization*. Next week you will evaluate the purity of the impure and purified vanillin samples by taking melting points. You will also determine the percent recovery.

1. Weigh into a 125 mL or 250 mL Erlenmeyer flask about 2 grams of your vanillin.

2. Add about 50 mL of water to the flask and stir vigorously.

3. Using a Bunsen burner, heat the solution just to the boiling point and stir until all the vanillin dissolves.

4. Place the flask in an ice bath for 5 minutes and stir occasionally.

5. Place a properly fitted piece of filter paper in your Buchner funnel. Make certain that all of the holes are covered. Wet the paper thoroughly with deionized water from your wash bottle and vacuum filter your vanillin crystals. Rinse the Erlenmeyer flask once or twice with 5 mL of ice-cooled water and use this water to wash the crystals in the funnel.

6. Empty the crystals onto a weighed piece of filter paper and place them in your desk for drying, being sure they cannot spill.

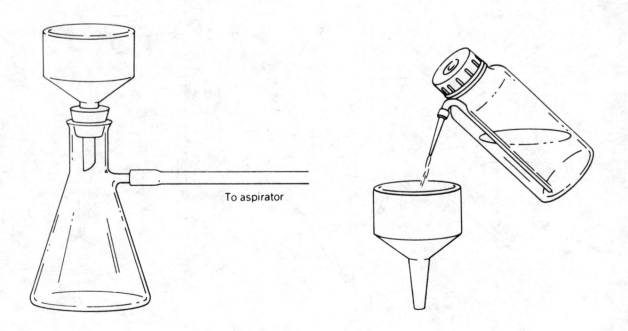

Figure 4-3

Prelaboratory Exercises - *Experiment 4*

For solutions to the starred problems, see *Appendix A*.

Name_____Date_____Lab Section_____

1. Define the following terms:

   a. evaporation

   b. saturated

   c. unsaturated

   d. homogeneous

   e. heterogeneous

2. Classify each of the following as element, compound, homogeneous mixture or heterogeneous mixture.

rust [iron(III) oxide]	_____	vinegar and oil	_____
diamond	_____	water (pure)	_____
carbon dioxide	_____	14 carat gold	_____
air in a sealed flask	_____	unsaturated aqueous sodium chloride	_____

3. Sodium chloride was selected for the evaporation part of this experiment for several reasons: sodium chloride is a solid at room temperature and is inexpensive, it has a low toxicity to humans and it has a water solubility that gives conveniently measured quantities. Explain why each of the following substances would not have been suitable for the evaporation experiment.

   a.* hydrogen chloride

   b.* calcium carbonate

   c.* arsenic(III) chloride

   d.   gold(III) chloride

   e.   sodium cyanide

   f.   sand (silicon dioxide)

   g.   ammonia

4. For the recrystallization of vanillin in this experiment, water was selected as the solvent for several reasons: water is inexpensive and has an extremely low toxicity, it has a reasonable boiling point and vanillin has a low solubility in cold water and a much higher solubility in hot water. Explain why water would not be suitable for the recrystallization of each of the following:

 a.* sand

 b. sodium chloride

 c. graphite

5. Barium chloride reacts with sodium sulfate in aqueous solution to give soluble sodium chloride and insoluble barium sulfate. What technique would you use to isolate the barium sulfate. Explain your answer.

6.* 8.5 mL of a sample of sea water solution was added to a 44.317 gram evaporating dish. The combination weighed 52.987 grams. After evaporation, the dish and contents weighed 44.599 grams.

 a. What is the mass percent of the salts in sea water?

 b. The sodium chloride content in the sea water sample was 2.69%. What percent error would have been caused if it had been assumed that the answer to 6-a above represented the percent of sodium chloride instead of the actual percent of all salts dissolved in the sea water (Caution: the answer is not 0.56%)?

    % error = $\dfrac{\text{difference}}{\text{theoretical}}$ × 100%

7. Recrystallization of a 1.75 gram sample of aspirin yielded 1.50 grams of aspirin. What was the percent recovery of aspirin?

Results and Discussion - *Experiment 4*
MIXTURES: SEPARATION AND PURIFICATION

Name_____Date_____Lab Section_____

A. EVAPORATION

1. Mass of evaporating dish _____

2. Mass of dish and sodium chloride solution _____

3. Mass of dish and sodium chloride (1st heating) _____

4. Mass of dish and sodium chloride (2nd heating) _____

5. Mass of dish and sodium chloride (3rd heating if necessary) _____

6. Mass of saturated sodium chloride solution _____

7. Mass of sodium chloride (residue in dish) _____

8. Mass percent of sodium chloride in saturated solution _____

9. Calculate the percentage deviation between your experimental
percent and the value in the solubility chart *(Appendix C)*
on page 399. [Hint: see *Prelaboratory Exercise 6-b*.) _____

10. Is sea water close to being a saturated sodium chloride solution?
Explain your answer. [Hint: see *Prelaboratory Exercise 6-b*.] _____

11. How could you tell that the original salt solution was saturated?

12. Why is the heat-cool-weigh cycle repeated until constant mass is attained?

B. FILTRATION

1. What did you observe when you mixed solutions of sodium carbonate and calcium chloride?

2. Why was gravity filtration used instead of evaporation?

3. What did you observe when you added HCl to the product of your reaction?

4. Did your observation in #B-3 support the conclusion that $CaCO_3$ was the product collected by filtration? Explain your answer.

5. Buildings and statues contain significant percentages of carbonates. Based on your observation in #B-3, what effects could acid rain have on these structures?

C. RECRYSTALLIZATION OF VANILLIN

1. Mass of crude vanillin + container   _____

2. Mass of container   _____

3. Mass of crude vanillin (also enter on page 75)   _____

4. Mass of paper (also enter on page 75)   _____

5. Why is the vanillin solution cooled in an ice bath before vacuum filtration?

6. How does the recrystallization procedure remove solvent soluble impurities?

7. What additional step could be added to this procedure to remove impurities insoluble in the solvent?

8. Gravity and vacuum filtration separate insoluble solids from a liquid phase. The choice depends on conditions. Suggest criteria you would apply to choose between them.

9. Suggest any ways you can think of to improve any part(s) of this experiment.

Experiment 5

MELTING POINTS

Learning Objectives

Upon completion of this experiment, students should have learned:
1. To take melting points using the capillary tube and cooling curve methods.
2. To graph a nonlinear phenomenon.
3. How to identify compounds and determine their purities, using melting points.

Text Topics

Physical properties, melting points, freezing point depressions (Malone, Chapters 2, 10)

Discussion

In last week's experiment you attempted to purify a sample of vanillin. Today you will determine the percent recovery and from a melting point measurement, the success of the purification. In addition, a second method of determining melting points will be studied.

PERCENT RECOVERY. Percent recovery and yield are extremely important from an economic perspective especially when industrial-sized amounts are used. The percent recovery is calculated by dividing the amount of the recrystallized sample by the mass of the starting solid and multiplying the result by 100%.

IDENTIFICATION TECHNIQUES. Anytime a compound is obtained from a natural or synthetic source, its identity must be determined and/or verified. Physical properties such as density and melting point are extremely useful for verification purposes because they involve quick and accurate procedures. These properties also provide a simple method of obtaining a rough indication of purity. A 5 Celsius degree difference between an experimental and an expected melting point for example, indicates that the sample is approximately 95% pure. Each 1% (up to about 10%), very roughly, depresses the melting point 1 Celsius degree. Impurities not only depress melting points but also broaden them. Thus, if compound A is supposed to melt at 65°C but experimentally it is found that heating causes it to begin softening at 57°C and totally liquify at 62°C, we report that the sample has a melting range of 57 - 62°C and is considerably impure.

The fact that contaminants tend to depress melting points enables one to distinguish between two compounds with the same melting point. Suppose one has two labeled bottles, *A* and *B*, and a third unlabeled bottle. The solids in all three bottles have the same melting point (say 65°C) within experimental error. If some of the unlabeled compound is mixed with *A* and with *B*, melting ranges of 64 - 65°C and 53 - 57°C respectively are observed. This strongly indicates that the unlabeled bottle is compound *A*. Why?

EXPERIMENTAL DETERMINATION OF THE MELTING RANGE. Two methods of melting range determination will be studied. In the first and routine method, a very small amount of sample in a capillary tube is heated and one observes when the first minute amount of liquid appears and when complete liquification occurs. the temperature for these two changes are recorded as the melting range (e.g., 91.4 - 93.0°C)

Another melting point technique that involves graphing a cooling curve is more time consuming and requires a much larger sample but potentially provides more information. This procedure involves the heating of the solid sample to about 10°C higher than its melting point. The liquid sample is allowed to cool slowly, solidify and then cool again. Temperatures are obtained at small time intervals and plotted against time on a graph. The principle can be most easily understood with an example.

If one had 10 grams of ice at -10°C and wanted to convert it to water at 10°C, using a constant energy input, the following would occur: It would take a short time to heat the ice from -10°C to 0°C. It takes much more energy to melt the ice than to change its temperature so it would now remain at 0°C for a substantial amount of time. Then it would take a short time to heat the water from 0°C to 10°C. The reverse process would occur if the water were cooled from 10°C to -10°C. Again, the temperature would remain constant for a long time while the water froze. The freezing point of a substance is the same as its melting point.

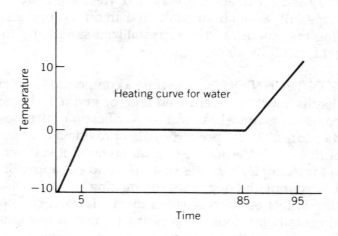

Figure 5-1

Procedure

A.   **PERCENT RECOVERY.**   Weigh the recrystallized vanillin from last week's experiment and determine its percent recovery.

B.   **CAPILLARY TUBE METHOD OF MELTING RANGE DETERMINATION.**   Using the technique described below, determine the melting ranges of samples *1, 2 and 3* below. [Hint: The melting ranges should occur between 60°C and 85°C.)   For the mixed sample, add approximately equal amounts of each compound to a mortar and grind them together with the pestle before inserting them into a capillary tube.

1. Crude vanillin from *Experiment 4.*
2. Recrystallized vanillin from *Experiment 4.*
3. ~50% recrystallized + ~50% phenyl carbonate (melting point ≈ 80°C)

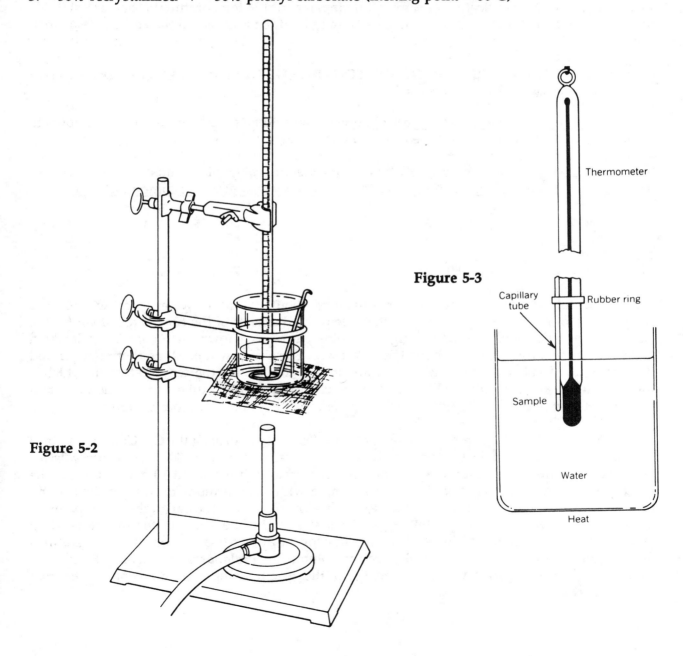

Figure 5-3

Figure 5-2

Refer to *Figures 5-2 and 5-3* for this discussion. Select a thermometer that covers the range -10°C to 110°C. Capillary tubes are very convenient sample holders as they are inexpensive, disposable and hold very small amounts of sample that can be easily observed during the melting process. Fill the capillary tube by pressing the open end onto the powdered sample until there is about a 0.5 - 1 cm length of sample in the tube. Now drop the capillary tube, sealed end down, through a 1 meter piece of 6 mm glass tubing that is being held on a hard surface. The impact of the capillary with the hard surface seldom results in breaking and causes the sample to drop to the bottom of the tube. Repeat the dropping procedure until the sample is packed in the bottom of the tube. Attach the capillary tube with a rubber ring (cut off a piece of rubber hose) to a thermometer with the sample even with the mercury bulb of the thermometer. Place the thermometer in a 250 mL beaker half full of water mounted above a Bunsen burner. Support the thermometer with a *split* rubber stopper and a clamp on a stand. Gently heat the water with *continual stirring* and observe the sample. As you approach the melting range, the heating rate should be very slow (around 2°C/min.) or large errors will be incurred. At the first indication of sample melting, record the temperature to the nearest 0.1°C. Continue to slowly heat until the sample has totally liquified and record the end of the melting range.

C. COOLING CURVE METHOD OF MELTING RANGE DETERMINATION. (Do either option 1 or option 2 at the discretion of the instructor.)

Option 1: Cooling curve of an Unknown. You will be given 1 gram of an unknown in a test tube (13×100 mm) that will be one of the following;

compound	melting range of high grade of compound (°C)	melting range of common commercial grade of compound (°C)
myristic acid	55 - 56	53 - 55
palmitic acid	63 - 64	60 - 62
stearic acid	69 - 70	65 - 67
phenyl carbonate	80 - 81	78 - 79

Clamp the test tube in a 150 mL beaker containing about 100 mL of water. You will have to wrap a strip of paper around the top of the tube if the diameter of the tube is too small for the clamp. Place the beaker on the ring and wire gauze as shown in *Figure 5-4*. Insert a thermometer into the tube containing the unknown. The clamp will hold the thermometer and tube upright. Heat the sample until it is about 15 - 20°C above the temperature at which it totally liquifies and shut off the burner. Lift the clamp holding the tube and thermometer to a height well above the beaker and reclamp it. Remove the beaker with beaker tongs.

As soon as possible, record the temperature (estimate to nearest 0.1°C). Continue to take temperature readings at 15 second intervals for the first minute and 30 second intervals for approximately the next 15 minutes (or until the temperature drops about 10 degrees below the solidifcation temperature.). *Stir* the cooling sample with the thermometer during the cooling process. Plot the temperature (vertical scale) vs. the time (horizontal scale) on the accompanying graph paper. Draw, with a straight edge, the best straight line you can for the descending temperatures before solidification and another line for the approximately constant temperatures during solidification. For an example, see the solution to *Prelaboratory Exercise #4* for this experiment. The melting point is read from the graph using the intercept of the two extrapolated lines.

Option 2: Cooling curve of water. Wash a 2 mL latex eyedropper bulb out thoroughly with deionized water. Fill it with deionized water and insert a thermometer into it until the end of the thermometer bulb just comes into contact with the end of the dropper bulb. Record the temperature to 0.1°C. Fill a 150 mL beaker with crushed ice and add 40 mL of ethylene glycol. Now immerse the bulb on the thermometer in the ice bath so that only the rim of the rubber bulb is above the liquid level and stir. Record the temperature every 30 seconds (vertical scale) vs. the time (horizontal scale) on the accompanying graph paper.

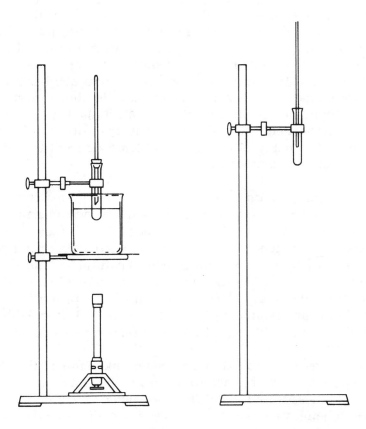

Figure 5-4

On a clear spring day, it might be nice to go $\underline{2[NaCl]_{(aq)}} \cdot$
$$C_7$$

vanillin  stearic acid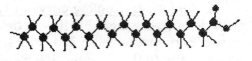

Chemical Capsule

Until at least 1828, organic chemistry was the chemistry of compounds derived from living things. It was believed that such chemicals had a special property called "vitalism" associated with them that differentiated these compounds from inorganic compounds. In 1828, Wöhler demonstrated that the organic compound, urea, could be made from inorganic chemicals. This led eventually to the redefining of organic chemistry to the chemistry of compounds that contain carbon. Consider then that we have one branch of chemistry for carbon compounds and another for all of the other elements. This is basically due to the importance of carbon in living things. Its importance is attributable to the fact that carbon is tetravalent (forms four bonds per carbon) and forms strong bonds to many other elements including itself (resulting in the formation of chains and polymers).

The compounds that will be used in this experiment are organic and all have practical uses. Vanillin occurs naturally in *vanilla* but is also synthetically produced from other chemicals. Contrary to the beliefs of people who preceded Wöhler and people today who buy "natural" products such as natural Vitamin C rather than the less expensive synthetic chemical, natural and synthetic chemicals are absolutely identical. Because of its pleasant aromatic odor and taste and relatively low toxicity, vanillin is frequently used as a flavoring agent in confectionery and other products.

The "fatty" acids that are possible unknowns in the cooling curve experiment are obtained from a reaction of fats (saponification reaction). The sodium salts of these long chain saturated acids are the main chemicals in soap. These salts can be considered to be composed of a long nonpolar organic chain that is attracted to relatively nonpolar grease and a polar part (the carboxylate group) that is attracted to the water. The soap lifts the grease off of clothes or dishes into the water.

Sailing, sailing over the seven seas.

Prelaboratory Exercises - *Experiment 5*

For solutions to the starred problems, see *Appendix A*.

Name_____Date_____Lab Section_____

1. A 2.5 gram sample of naphthalene (melting range 73.2 - 77.1°C) is recrystallized and 1.8 grams of purified material (melting range 79.3 - 80.1°C) are recovered.

    a.* What is the percent recovery?   _____

    b. What conclusions can you draw from the melting ranges?

2.* Cinnamic acid and urea melt at 133°C. The melting point of an unknown is found to be 133°C. When the unknown is mixed with urea, the melting range of the mixture is 120° - 125°C. Is it possible to identify the sample from this information? If not, what further tests would you perform?

3. Virtually all compounds with ionic bonds (metal - nonmetal, metal - polyatomic ion) are solids at room temperature and most have melting points above 200°C. Compounds that contain only nonmetals have covalent bonds (called molecular compounds) exist as gases, liquids and solids at room temperature. Rank the following compounds from highest melting point to lowest.

    helium, carbon dioxide, sodium chloride, nitrogen, water

4.*   A sample of reagent grade biphenyl was heated approximately 25° above its liquification temperature. The sample was then allowed to cool. The temperature was recorded during cooling as a function of time and the data below was collected. Plot the temperature on the y axis and the time on the x axis and determine the melting point of biphenyl.

time (min.)	temp. (°C)	time (min.)	temp. (°C)	time (min.)	temp. (°C)
0.00	96.0	3.50	70.1	9.00	69.2
0.50	88.9	4.00	70.0	10.00	68.1
1.00	83.7	5.00	70.0	11.00	65.8
1.50	78.0	6.00	70.0	11.50	63.5
2.00	73.9	7.00	70.0	12.00	59.0
2.50	70.8	8.00	69.9	12.50	56.0
3.00	70.0				

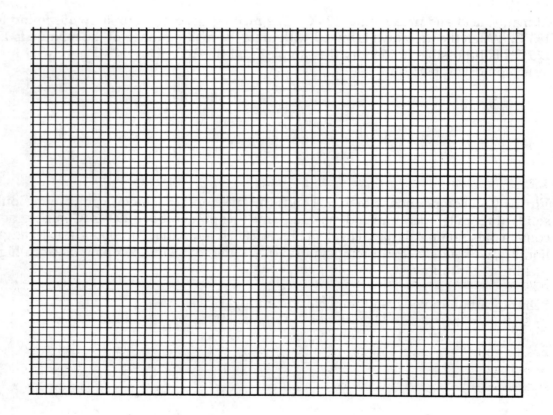

Results and Discussion - *Experiment 5*
MELTING POINTS

Name_____Date_____Lab Section_____

A. PERCENT RECOVERY.

1. Mass of weighing paper (from *Experiment 4*)   _____

2. Mass of weighing paper + recrystallized vanillin   _____

3. Mass of recrystallized vanillin   _____

4. Mass of crude vanillin (from *Experiment 4*)   _____

5. Percent recovery   _____

6. What are the major factors that you think contributed to loss of part of the sample?

B. CAPILLARY TUBE METHOD OF MELTING RANGE DETERMINATION.

<u>melting range (°C)</u>

1. Crude vanillin from *Experiment 4*   _____

2. Recrystallized vanillin from *Experiment 4*   _____

3. 50% recrystallized vanillin + 50% phenyl carbonate (mp = 80°C)   _____

4. Basing your answer on a literature melting point of 81-82°C for vanillin and your results to *B-1* and *B-2*, comment on the purity of the crude and recrystallized vanillin and the success of your recrystallization.

C. COOLING CURVE METHOD OF MELTING RANGE DETERMINATION.

time (min.)	temp. (°C)	time (min.)	temp. (°C)	time (min.)	temp. (°C)
0	_____	4.5	_____	10.0	_____
0.25	_____	5.0	_____	10.5	_____
0.50	_____	5.5	_____	11.0	_____
0.75	_____	6.0	_____	11.5	_____
1.0	_____	6.5	_____	12.0	_____
1.5	_____	7.0	_____	12.5	_____
2.0	_____	7.5	_____	13.0	_____
2.5	_____	8.0	_____	13.5	_____
3.0	_____	8.5	_____	14.0	_____
3.5	_____	9.0	_____	14.5	_____
4.0	_____	9.5	_____	15.0	_____

On the accompanying piece of graph paper, plot time on the x axis and temperature on the y axis.

1.[1] Unknown number _____

2.   What is the melting point (or range) of your sample? _____

3.[1] What is the name of your unknown _____

4.   Suggest any ways you can think of to improve any part(s) of this experiment.

---
[1]Answer only if option 1 was performed.

Experiment 6

PERIODIC PROPERTIES, SOLUBILITY AND EXTRACTION

Learning Objectives

Upon completion of this experiment, students should have learned:
1. The concept of electronegativity and its relationship to bond and molecular polarity.
2. The relationship of molecular polarity to solubility.
3. The concept of extraction.
4. The colors of colored ions.

Text Topics

Electronegativity, polarity, solubility, periodicity (Malone, Chapters 5, 6)

Comments

This is a relatively short experiment and can be performed in the same lab period as *Experiment 7*. For additional problem solving experience involving periodic properties, and bond and molecular polarities, see *Exercises 11, 12*.

Discussion

The periodic table. Take a look at all the substances around you. There are solids, liquids and gases of different colors and densities. Is there any relationship between any of these substances? In the last century, chemists found a number of relationships among different kinds of matter. One of the discoveries that brought order out of apparent chaos was that the properties of the elements are periodic. In 1869, Dimitri Mendeleev arrived at an early version of the periodic table by arranging the elements in terms of increasing atomic mass (now correctly arranged by atomic number) and starting new groups whenever vertical alignment of elements of similar properties resulted. Theoretical models of the electronic structure of the atom, developed more than a half century after Mendeleev's ingenious arrangement, are completely consistent with the Mendeleev table.

On the space of less than one page, the periodic table provides explanations for many physical and chemical properties of the elements. Predictions of chemical formulas (NaCl, CaCl$_2$, AlCl$_3$) and understanding of trends of metallic and nonmetallic behavior, atomic radii, ionization energies and electron affinities are easily obtained from the periodic table.

Another periodic property, **electronegativity**, can be utilized to qualitatively predict solubility, a property intimately related to today's experiment. The concept of electronegativity was first developed by the only person to have ever won two unshared Nobel Prizes, most astoundingly in two different fields (chemistry and peace), Linus Pauling. The electronegativity scale is a measure of the ability of an atom of an element to attract electrons to itself in a chemical bond. Pauling's scale still finds very common usage today and will help us to explain our observations today on solubility.

Fluorine is the most electronegative element and the electronegativity decreases going to the left and/or down the periodic table from fluorine. A few of the most commonly used electronegativity values are in the table below.

H = 2.1

C = 2.5	*N = 3.0*	*O = 3.5*	*F = 4.0*
	P = 2.1	*S = 2.5*	*Cl = 3.0*
			Br = 2.8
			I = 2.5

When two elements bonded together have different electronegativities, the electrons in the bond will spend more time in the vicinity of the more electronegative atom. This results in a polar bond with each atom bearing a partial charge. In cases where the electronegativity differences are large (>1.7) such as for many metal - nonmetal bonds, the electrons spend most of their time on the more electronegative atom and the bond is termed ionic. If the two bonded atoms have identical or very similar electronegativities (difference <0.5), the electrons are shared almost equally and the bond is called covalent or nonpolar covalent. Electronegativity differences between 0.5 and 1.7 lead to the most interesting bonds. In these cases the electrons are shared but not equally and the bond is called a polar covalent bond. For example, in HCl, the chlorine atom has an electronegativity that is 0.9 greater than that of the hydrogen. For this reason, HCl is a polar molecule with a dipole pointing from the hydrogen toward the chlorine.

$$\overset{\delta+}{H} \underset{\longrightarrow}{\hspace{2em}} \overset{\delta-}{Cl} \qquad \cdots \overset{\delta+}{H}-\overset{\delta-}{Cl}\cdots\overset{\delta+}{H}-\overset{\delta-}{Cl}\cdots\overset{\delta+}{H}-\overset{\delta-}{Cl}\cdots\overset{\delta+}{H}-\overset{\delta-}{Cl}\cdots$$

The intermolecular attractions that result from the opposite charges help to hold molecules near each other and the result is that more energy is needed to convert a liquid into a gas. Because of the strong intermolecular hydrogen bonds in water, the boiling point of water is much higher than would be expected in the absence of these intermolecular dipole - dipole attractions.

Solubility. Molecular polarity also has a significant effect on solubility. For a solute to dissolve in a solvent, the solute molecules (or ions if the compound is ionic) must be separated. Dissolving will only occur if the interaction of the solute with the solvent is energetically favorable. You know that sugar dissolves in water but gasoline does not. Sugar molecules can form strong intermolecular hydrogen bonds to water and the overall energetics are favorable for dissolving. Gasoline on the other hand does not have polar bonds and is not attracted to water.

Gasoline molecules are "happier" next to each other than next to water, while water molecules are "happier" next to each other than next to gasoline molecules.

β-D-glucose

2,2,4-trimethyloctane
(isooctane)

While bond polarities can easily be estimated from electronegativity differences, determination of molecular polarity is a more complicated matter. For a molecule to be polar, it must have one or more polar bonds. Thus gasoline, with all nonpolar bonds must be nonpolar regardless of other considerations. A molecule is only polar when it has polar bonds and a shape that does not cause cancellation of the bond dipoles. Although carbon dioxide has polar bonds, the molecule is nonpolar because of its linear shape that results in cancellation of its bond dipoles. Water on the other hand has a bent shape and the bond dipoles add to give the molecule a resulting dipole. We encounter an experimental result of this almost everyday when we open a bottle of soda and watch the bubbles of CO_2 escape from the water. The nonpolar carbon dioxide is not substantially soluble (0.16 g CO_2/100 mL water at 20°C) in the polar water and thus escapes from it as soon as the container is opened and the pressure is lowered.

While molecular polarities can be determined theoretically from electronegativities of component atoms and molecular shapes, they can also be determined experimentally from a study of the miscibilities of liquids (when discussing the solubility of one liquid in another, the term miscibility is often used). For this analysis, remember that molecules with similar polarities generally attract each other and often dissolve in each other. Molecules with dissimilar polarities are usually relatively insoluble in each other. These observations lead to the general rule for solubility "like dissolves like".

Extraction. Solubility also plays a key role in an extremely important separation technique called extraction. Consider an immiscible pair of liquids such as vinegar (polar - 95% water, 5% acetic acid) and oil (nonpolar). Suppose the vinegar has a small amount of a nonpolar solute dissolved in it and you shake the vinegar with some oil. The solute will be given a choice. As its polarity is closer to that of the oil than of vinegar, the solute will probably shift to the oil. We say that the solute has been extracted from the vinegar by the oil. If the solute had been polar, it would have remained in the vinegar.

Colored ions. Another property that will be qualitatively investigated today is the color of ions. Recall that ions are formed by loss of electrons (positive ions - cations) or gain of electrons (negative ions - anions). These electrons are lost or acquired when the element enters into bonding with another element of significantly different electronegativity (> 1.7). One of the goals will be to determine if there is any periodicity to the color of ions. To do this, you will study dilute solutions of compounds which have ionic bonds. When the compounds dissolve in water, dissociation into cations and anions occurs. Ions can be colored or colorless and generally the color of an ion does not depend on its partner [sodium chloride and sodium nitrate solutions are colorless but copper(II) sulfate and copper(II) nitrate solutions are blue because sodium, chloride, and nitrate ions are colorless and copper(II) is a blue ion]. If we had solutions available for you containing all of the commonly encountered ions, you would notice that most of the common ions are colorless. *This makes it very useful for identification purposes for you to observe and remember the names and colors of the colored ions.* Often you can conclude upon visual inspection that an unknown blue solution possibly contains copper(II) ions.

Procedure

A. Miscibility.

Pour the following liquids into test tubes and attempt to mix. Report your observations.

1. 5 mL water + 2 mL ethanol (CH_3CH_2OH)
2. 5 mL water + 2 mL 1-butanol ($CH_3CH_2CH_2CH_2OH$)
3. 5 mL water + 2 mL 1-hexanol ($CH_3CH_2CH_2CH_2CH_2CH_2OH$)
4. 5 mL water + 2 mL kerosene (hydrocarbon with 10 to 16 carbons)

Predict what will happen for the next two mixtures and then test your predictions.

5. 5 mL ethanol + 2 mL kerosene
6. 5 mL hexanol + 2 mL kerosene

B. Extraction.

1. Consider the polarities of water and elemental iodine. Should iodine be very soluble in water? Visually inspect the stock solution of aqueous iodine. Can you tell if you were right? Look up the solubility of iodine in water in the *Handbook of Chemistry and Physics*. Do you think iodine should be very soluble in kerosene? Will the iodine prefer water or kerosene if you give the iodine a choice? Transfer 5 mL of the aqueous iodine solution to a test tube, add 2 mL of kerosene, stopper the tube and shake vigorously. Record your observations and conclusions.

2. Methylene blue is an ionic compound (and therefore polar) but it is a very large molecule and therefore has limited solubility in water. Will the methylene blue prefer water or kerosene? Pour 5 mL of the aqueous methylene blue solution into a test tube, add 2 mL of kerosene, stopper the tube and shake vigorously. Record your observations and conclusions.

C. Colored ions.

Four sets of samples will be provided in sealed vials. The first two sets have been designed to enable you to determine the effect of the cation or anion on the color of its partner. For the third set, you should focus your attention on the colors of the cations. Assume that the anions (either chloride or nitrate) are colorless and look for a correlation between color and position in the periodic chart. For the fourth set, focus your attention on the colors of the anions. 0.1 M solutions of the compounds listed below will be available for your observation.

Set 1: 0.1 M solutions of copper(II) acetate, copper(II) nitrate, copper(II) sulfate

Set 2: 0.1 M solutions of ammonium chromate, potassium chromate, sodium chromate

Set 3:

aluminum nitrate	manganese(II) chloride
barium chloride	mercury(I) nitrate
calcium chloride	mercury(II) nitrate
cerium(III) nitrate	nickel(II) chloride
chromium(III) chloride	potassium chloride
cobalt(II) chloride	silver nitrate
copper(II) chloride	sodium chloride
iron(III) chloride	strontium chloride
lead(II) nitrate	tin(II) chloride
lithium chloride	tin(IV) chloride
magnesium chloride	zinc nitrate

Set 4:

sodium acetate	sodium iodate
sodium bromide	sodium iodide
sodium carbonate	sodium nitrate
sodium chlorate	sodium oxalate
sodium chloride	potassium permanganate
sodium chromate	sodium phosphate
sodium dichromate	sodium sulfate
potassium ferricyanide	sodium sulfite
potassium ferrocyanide	sodium thiocyanate
sodium hydroxide	sodium thiosulfate

84

Chemical Capsule

Isotopes are atoms of the same element with different numbers of neutrons. Nuclear stability depends on the proton to neutron ratio and determines if an atom is radioactive and if so, the type of radioactive decay. Chemical reactivity does not depend significantly on the proton to neutron ratio although there are small but measurable effects on reaction rates. One notable exception involves the isotopes of hydrogen. These isotopes and their differences are important enough for them to earn their own names. 99.98% of the naturally occurring hydrogen atoms are $_1^1H$ and their nuclei consist simply of 1 proton. The remaining 0.02% of hydrogen atoms also have a stable nucleus consisting of 1 proton and 1 neutron. $_1^2H$ is called deuterium. A third isotope of hydrogen, tritium ($_1^3H$), has 1 proton and 2 neutrons and undergoes beta decay with a half life of 12.5 years.

$$_1^3H \rightarrow _2^3He + _{-1}^0e$$

If a chemical reaction involves the breaking of a bond between an atom such as carbon or oxygen and hydrogen, the reaction proceeds about seven times slower with $_1^2H$ than with $_1^1H$. This means that drinking 100% D_2O instead of H_2O would considerably disrupt the balance of reactions in your body and possibly cause death. The difference in chemical behavior of the hydrogen isotopes should not be surprising considering the fact that the addition of a neutron to $_1^1H$ doubles its mass. For other elements, the mass differences are usually small percentages of the total mass.

	H_2O	D_2O
boiling point (°C)	100.00	101.42
melting point (°C)	0.00	3.81
density (g/mL) (25°C)	0.9970	1.1044

One of the potential solutions to society's future energy needs is to develop the technology necessary to make nuclear fusion commercially economical. Part of the fuel for nuclear fusion will probably be deuterium. Although deuterium only makes up 0.02% of the naturally occurring hydrogen atoms, 0.02% of all the hydrogen in the oceans would provide enough fuel to satisfy our energy needs for over a million years.

Prelaboratory Exercises - *Experiment 6*

For solutions to the starred problems, see *Appendix A*. For additional problem solving experience involving periodic properties and bond and molecular polarities, see *Exercises 11, 12*.

Name_____Date_____Lab Section_____

1. Which of the following pairs would you expect to have a higher boiling point? Explain your answers.

   a.* $CH_3CH_2CH_3$ or CH_3CH_2OH

   b. CH_3OCH_3 or CH_3CH_2OH

   c.   HF or HCl

   d.* CH_3F or CH_3Cl

2. Should the first compound listed be more soluble in the first solvent or the second? Explain your answer.

   a.* NaCl in water or kerosene

   b.   KOH in water or kerosene

   c.* HCl in water or kerosene

   d.   wax in water or kerosene

3. The chart below gives the formulas of compounds of elements in periods 2 and 3 with chlorine for groups IA through IVA (1, 2, 13, 14) and with hydrogen for groups VA through VIIA (15 - 17).

LiCl	$BeCl_2$	BCl_3	CCl_4	NH_3	H_2O	HF
NaCl	$MgCl_2$	$AlCl_3$	$SiCl_4$	PH_3	H_2S	HCl

a. Describe any trends or periodicity in the formulas.

b. Chemists use a bookkeeping system called oxidation numbers to keep track of electrons in reactions. While the guidelines for the determination of oxidation numbers are somewhat complex, the oxidation numbers of the elements in the compounds above can be determined rather simply by applying the following rules: the sum of the oxidation numbers in a compound must be zero, the oxidation number of chlorine in a binary compound is -1 and the oxidation number of hydrogen when bonded to a non metal is +1. Using $FeCl_3$ as an example, the sum of the oxidation numbers of the iron and three chlorines must be zero. According to the rules given, chlorine has an oxidation number of -1 thus it is possible to calculate the oxidation number of iron to be 3 [e.g., x + 3(-1) = 0, x = 3]. From the formulas given above and the rules given, calculate the oxidation numbers of the elements in the chart below.

Li ____ Be ____ B ____ C ____ N ____ O ____ F ____

Na ____ Mg ____ Al ____ S ____ P ____ S ____ Cl ____

c. Describe any trends or periodicity in the oxidation numbers.

d. Do the oxidation numbers correlate with group numbers (American system, IA - VIIA)? Explain you answer.

Results and Discussion - *Experiment 6*
PERIODIC PROPERTIES, SOLUBILITY AND EXTRACTION

Name_____Date_____Lab Section_____

A. Miscibility (for miscibility, use M for miscible, P for partially miscible and I for immiscible)

mixture	observations	miscibility
1. water + ethanol		___
2. water + butanol		___
3. water + hexanol		___
4. water + kerosene		___

	prediction	observations	
5. ethanol + kerosene			___
6. hexanol + kerosene			___

B. Extraction

1. Should iodine be very soluble in water? Explain your answer. _____

2. Solubility of iodine in water from *Handbook of Chemistry and Physics*. _____
   edition____page____

3. Should iodine be very soluble in kerosene? _____
   Explain your answer.

4. Observations when aqueous iodine and kerosene are mixed.

   Explanations and conclusions.

5. Discuss your observations about the color of iodine in water and kerosene and suggest an explanation for your observations.

6. Observations when aqueous methylene blue and kerosene are mixed.

   Explanations and conclusions.

C. Colors.

1. Does the anion seem to have any effect on the color of the cation?   _____
    Explain your answer.

2. Does the cation seem to have any effect on the color of the anion?   _____
    Explain your answer.

3. List the colored cations and their colors.

cation	color	cation	color	cation	color
_____	_____	_____	_____	_____	_____
_____	_____	_____	_____	_____	_____

4. Can you make any generalizations about color versus position in the periodic chart?

5. List the colored anions and their colors.

anion	color	anion	color	anion	color
_____	_____	_____	_____	_____	_____
_____	_____	_____	_____	_____	_____

6. Suggest any ways you can think of to improve any part(s) of this experiment.

Experiment 7

LEWIS STRUCTURES AND MOLECULAR MODELS

Learning Objectives

Upon completion of this experiment, students should have learned:
1. How to draw Lewis structures of simple molecules and polyatomic ions.
2. How to construct models of simple molecules and polyatomic ions.

Text Topics

Lewis structures and geometry of molecules and polyatomic ions (Malone, Chapter 7)

Comments

This is a relatively short experiment and can be performed in the same laboratory period as *Experiment 6*. **Considerable time is saved and the educational value of the experiment is increased if the Lewis structures are drawn as a prelaboratory exercise.**

Discussion

Although it has recently become possible to image molecules and even atoms using a scanning tunneling microscope, most of our information about molecular structure comes from interpretation of physical, chemical and spectroscopic properties of substances. This information often enables us to piece together a 3-dimensional picture or model of the molecule. On paper, one of the best methods we have of representing this model is by drawing a Lewis structure of the molecule or ion. The ability to draw Lewis structures for covalently bonded compounds and polyatomic ions is essential for the understanding of polarity, resonance structures, chemical reactivity and isomerism. Molecular models are a useful tool to help you visualize structures, especially when the ion or molecule is not planar. It is not the intention here to teach you all aspects of drawing Lewis structures or construction of molecular models. The goal here is to supplement your text by guiding you through some examples and providing a few insights and tactics where difficulty is often encountered. Be sure to refer to your chemistry textbook for more complete instruction and details.

Lewis structures. Remember the absolute rule that the Lewis structure must show the correct number of electrons. For a molecule, the sum of the valence electrons is the correct number. For a positive ion, subtract one electron for each positive charge from the number of valence electrons. For a negative ion, add one electron for each negative charge to the number of valence electrons. For example, formaldehyde, CH_2O, must show $4 + 2 + 6 = 12$ electrons and nitrite, NO_2^-, must show $5 + 6 + 6 + 1 = 18$ electrons.

Many schemes have been devised to help with the drawing of Lewis structures but the following is one of the simplest. Make a list of the elements present (if more than one atom of the element is present in the molecule, repeat the name in the list for each atom). Next to the list, make two columns, the first for valence electrons and the second the number of electrons needed for the atom to complete its octet (or duet for hydrogen). The following example will illustrate the technique for formaldehyde (CH_2O):

Elements	Valence electrons	# needed
C	4	4
H	1	1
H	1	1
O	6	2
	12	8 8/2 = 4 bonds

After listing the elements, valence electrons and the number of electrons needed, find the total number of valence electrons (12 for formaldehyde) and electrons needed (8 for formaldehyde). For formaldehyde, 12 and only 12 electrons must be showing in the completed Lewis structure. As there are two electrons per bond, the number of electrons needed is divided by 2 to determine the number of bonds in the molecule. Why do this? Once you gain experience, you will simply write down the number of elements and fill in the total number of electrons while satisfying the duet and octet rules. But to begin with, this technique enables you to ascertain the right number of bonds and avoid some random guessing.

Now we must consider the sequence of bonding. The three possible choices are:

$$\begin{array}{ccc} & C & O \\ H\ C\ O\ H & H\ O\ H & H\ C\ H \end{array}$$

When the atoms can link together in different ways, the symmetrical choice is usually correct. Because the first structure has lower symmetry than the other two, we could correctly eliminate it on this basis. We will leave it in for this discussion and continue by inserting the 4 bonds (8 e^-) indicated by our earlier calculation and adding the remaining 4 e^- to satisfy the octet rule.

Another method of deciding among different structures involves the use of formal charges on atoms.

formal charge = valence electrons - bonds - nonbonded electrons

Different structures usually have a different distribution of formal charges and usually the structure with the minimal number of formal charges will be correct.

The three structures are technically correct Lewis structures for CH_2O but only one correctly represents formaldehyde. The minimization of formal charges enables us to strongly favor the third and correct structure.

For ions, the procedure is similar but a modification is needed. In the column for valence electrons, add in the negative of the charge. If the ion has a negative charge, it has extra electrons that it has acquired from its partners. In the number needed column enter the actual charge. The following example will illustrate the technique for nitrite, NO_2^-.

Elements	Valence electrons	# needed
N	5	3
O	6	2
O	6	2
e^-	1	-1
	18	6 6/2 = 3 bonds

Use of the symmetry guideline leads to the sequence ONO rather than OON and in addition, a formal charge analysis (try it!) also favors ONO. The double headed arrow below indicates that the two Lewis structures are resonance structures that differ only by the position of electrons.

When resonance structures can be drawn, none of the Lewis structures correctly depicts the structure but one must try to imagine a hybrid (or enhanced average) as a better model for the species. One final note about the method introduced above - if the number of valence electrons comes out to be an odd number, the best Lewis structure you will be able to draw will not satisfy the octet rule.

Finally, we turn to geometry. Once the Lewis structure is drawn, it is possible to apply either VSEPR (valence shell electron pair repulsion) theory or a hybridization model to predict the shape of the species. Both systems when appropriately used will predict with very few exceptions the same shape. The table below summarizes the theories.

groups[1] of electrons around central atom	shape	bond angles	hybridization
2	linear	180°	sp
3	planar	120°	sp^2
4	tetrahedral	109.47°	sp^3

[1]The number of groups of electrons is equal to the sum of the number of neighbor atoms and nonbonded electron pairs.

For the formaldehyde molecule drawn earlier, there are 3 groups of electrons around the central carbon. Notice the double bond counts as 1 group of electrons and not 2! The 3 groups result in a prediction that the atoms around carbon are 120° apart in a plane.

For nitrite, there are also 3 groups of electrons around the central nitrogen as the nonbonded electron pair counts as a group. Again VSEPR theory predicts 120° bond angles and a bent ion. When 4 groups of electrons surround the central atom, a tetrahedral structure results which we cannot easily represent on paper except as a projection. While projections are occasionally used, CH_4 is often written in planar form with bond angles that appear to be 90°. You should recognize when you see the planar drawing that the bond angles are actually 109.5° and that the molecule is tetrahedral.

$$H - C - H$$

with H above and below C.

Molecular polarity. Once the Lewis structure has been drawn and the geometry resolved, it is possible to determine if the molecule is polar. A knowledge of polarity is extremely useful for predicting relative boiling points, solubility, and chemical reactivity. For a molecule to be polar, it first must have polar covalent bonds and then it must have a geometry that does not result in the cancellation of bond dipoles. Except for carbon - hydrogen bonds, almost all bonds between different nonmetals are polar covalent. Thus if a molecule contains two different nonmetals and lacks the symmetry necessary for cancellation of bond dipoles, the molecule will be polar.

Molecular models. Most molecular model kits will help you visualize the structures of molecules in three dimensions. However, the most common ball and stick models give slightly incorrect bond angles when multiple bonds are present. For instance, for formaldehyde, the carbon used will have its holes in a tetrahedral arrangement and the H-C-H and H-C-O bond angles appear to be 109.5° and 125.3° respectively. The correct bond angles are very close to 120°. Also the model will incorrectly show the double bond to consist of two identical bent bonds when actually the double bond is composed of two different kinds of bonds; a σ bond and a π bond.

Procedure

Recognizing the discrepancies discussed above, draw Lewis structures and construct models for the molecules and polyatomic ions listed in the *Results and Discussion* section.

Results and Discussion - *Experiment 7*
LEWIS STRUCTURES AND MOLECULAR MODELS

Name_____Date_____Lab Section_____

For each molecule in the chart below, draw a Lewis structure, construct a model, determine the bond angle(s) and the molecular polarity (P = polar, N = nonpolar). Note that elements through the second period (Ne) virtually never exceed the octet rule. Also hydrogen has one and only bond and carbon almost always has four bonds (except for carbon monoxide and cyanide ion).

Molecule	Lewis Structure	Bond Angle	Polarity
F_2			____
N_2			____
ICl			____
CO_2		____	____
H_2O		____	____
H_2O_2		____	____

96

Molecule	Lewis Structure	Bond Angle	Polarity
SO_2		____	____
NH_3		____	____
N_2H_4		____ H-N-H	____
		____ H-N-N	
CH_4		____	____
C_2H_6		____ H-C-H	____
		____ H-C-C	
C_2H_4		____ H-C-H	____
		____ H-C-C	
C_2H_2		____	____
HCN		____	____

For each polyatomic ion in the chart below, draw all the reasonable resonance structures and construct a model of one of the resonance structures of each ion. Determine the indicated bond angle(s).

Polyatomic Ion	Lewis Structures	Bond Angle
OH⁻		
CN⁻		
N_3^-		____
ClO_2^-		____
ClO_3^-		____
NO_3^-		____

98

Polyatomic Ion	Lewis Structures	Bond Angle
CO_3^{2-}		____
SO_3^{2-}		____
NO_2^+		____

Optional: For the formulas N_2O and CH_4O, two bond sequences are possible that enable the rules of Lewis structures to be followed (correct number of valence electrons and octet rule). Draw Lewis structures of the two structures for each. Based on formal charges circle the preferred and in this case the correct structure. Fill in the bond angles for the correct structure.

N_2O ____

CH_4O

____
H-C-H

____
H-C-O

____
C-O-H

Experiment 8

PAPER CHROMATOGRAPHY

Learning Objectives

Upon completion of this experiment, students should have learned:
1. Principles and applications of chromatography.
2. To identify components of an unknown mixture, using paper chromatography.

Text Topics

Separation and purification (Malone, Chapter 2)

Discussion

One of the most important aspects of chemistry is the analysis of mixtures. Usually, a mixture must be separated into pure compounds before chemical analysis can be attempted. Polluted water often contains several pesticides as well as chemicals such as PCBs and analytic techniques cannot identify one pesticide (e.g., DDT) when another is present unless a separation is first performed. Many mixtures are separated using some variation of a technique discovered by Michael Tswett (Russian botanist) early in the 20th century.

Tswett allowed a mixture of pigments extracted from plants to percolate down through a column of calcium carbonate. Solvent was added from the top as needed to cause continuous movement of the pigments down the column. Pigments that were more strongly attracted to the stationary phase (calcium carbonate) and had less affinity for the solvent, moved down the column more slowly than pigments that had greater affinity for the solvent and had weaker attraction for the stationary phase. Tswett observed that the pigments had separated into several differently colored bands as a result of the fact that they moved down the column at different rates. The term "chromatography" was coined to describe the phenomenon.

The paper chromatography experiments you will perform today utilize the same principles Tswett employed. A piece of paper, spotted with pure compounds and mixtures, is placed in a solvent as illustrated. Assume spots 1, 2, and 3 are pure compounds and spot 4 is an unknown mixture of the compounds. The solvent (moving phase) will move up the paper (stationary phase). When the solvent reaches the spots, the components of each spot have

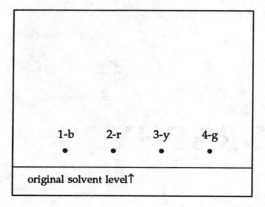

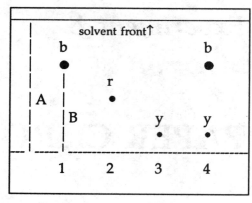

$b = blue$ $r = red$ $y = yellow$ $g = green$

Figure 8-1   **Figure 8-2**

choices. They can dissolve in the solvent and progress up the paper or they can stay adsorbed on the paper and resist movement. The choice depends on several factors including the polarities of the compounds, the solvent and the paper. The solvent and the stationary phase are selected so that the components spend some time in each phase and do not move right along with the solvent front or stay at the origin. If the solvent and the stationary phase are selected properly, different compounds will move up the paper at different rates and separate.

The spots of compounds that are less strongly adsorbed on the paper will move up the paper faster than the spots of the more strongly adsorbed compounds. When the solvent front nears the top of the paper, the chromatogram is removed from the solvent and the solvent front marked with a pencil. To find out the composition of the fourth spot, two factors are considered, color and relative distance moved. To quantify the relative distances, the R_f (ratio to front) value of each spot is calculated.

$$R_f = \frac{\text{distance from origin to center of spot}}{\text{distance from origin to solvent front}} = \frac{B}{A} \text{ for spot 1}$$

In the chromatogram on the top part of this page, it can be seen that spot 4 (the green spot) has compounds with the same color and R_f values as pure compounds 1 and 3. This provides strong evidence (but not proof) that spot 4 contains compounds 1 and 3 and does not contain compound 2. Paper chromatography then serves as a separation technique and can also assist with identification if the possible compounds in a mixture are available in pure form.

Procedure

Paper chromatography. The same general procedure will be followed for your two chromatograms. Following *Figures 8-3* and *8-5*, draw straight pencil lines two centimeters from the bottoms of both pieces of chromatography paper and place pencil dots at equal intervals along the lines. Spot the larger paper with the felt-tip pens, as shown in the diagrams. The spots should be about this size ●.

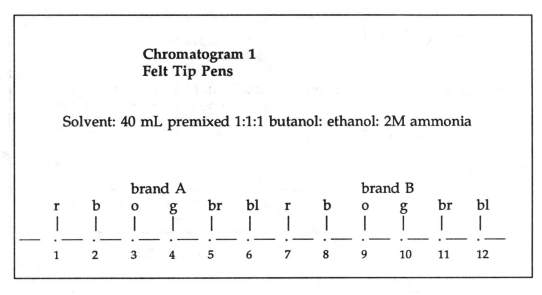

Figure 8-3

After the spotting (*Figure 8-3*) is complete, roll the paper into a cylinder and staple it so that there is a small gap between the two ends (*Figure 8-4*). The ends of the paper should **not** overlap. Add 40 mL of the butanol, ethanol, ammonia solvent to a 600 mL beaker. Gently put the paper cylinder (spotted edge down) into the beaker and cover the beaker with plastic wrap. Do not move or turn the beaker again. At this point, you should begin your second chromatogram but keep your eye on the first one. When the solvent front reaches about 2 cm below the top of the paper, remove the chromatogram. Mark the solvent front with a pencil and dry the paper in the hood, with a hot air blower.

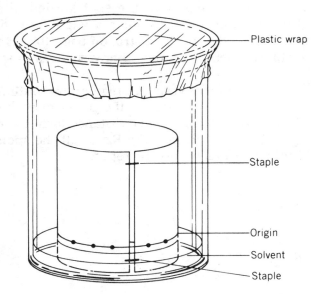

Figure 8-4

Following *Figure 8-5*, spot the smaller piece of paper with the metal ion solutions and with your unknown. To do this, draw some of the solution up into a capillary tube and apply it to the paper on the appropriate pencil dot. Again, the spots should be about this size, ●. Before spotting your chromatogram, perfect your spotting techniques on a "practice" piece of chromatography paper. Put 7 mL of 6 M hydrochloric acid and 25 mL of acetone into your 400 mL beaker and stir the mixture until the liquids are *thoroughly mixed*. Staple the paper into the form of a cylinder, as you did with the larger paper; gently place it into the 400 ml beaker and cover the beaker with the plastic wrap. When the solvent front reaches about 1 cm below the top of the paper, remove the paper from the beaker and mark the solvent front with a pencil.

Dry the metal ion chromatogram with a hot air blower, outline and record colors of any spots that are visible, place it over a petri dish **in the hood** containing concentrated ammonia *[Caution: avoid touching the ammonia or breathing the vapors]* and cover with a watch glass

1 - iron(III)
2 - copper(II)
3 - cobalt(II)
4 - manganese((II)
5 - mixture of 1, 2, 3, 4
6 - unknown

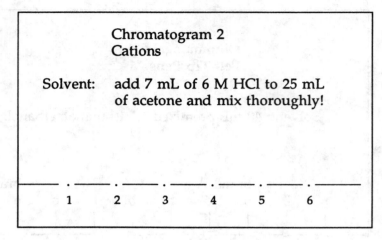

Figure 8-5

for a few minutes until the spot labeled Mn^{2+} has appeared. Be sure to expose the entire chromatogram to the ammonia vapor. Outline any new spots that appear and record any changes in color in the matrix on page 109. Heat the chromatogram a second time with the hot air blower and again record any changes in spot color.

On both of your chromatograms, outline each of the spots with a pencil and measure the distance from the origin to the center of each spot. The distance from the origin to the center of the spot, divided by the distance form the origin to the solvent front (see *Figure 8-2*), is the R_f of the spot. Record the R_fs of all the spots on your data sheet and turn in your finished chromatograms with your report.

Chromatogram Summary

Description	Chromatogram 1 Felt-Tip Pens	Chromatogram 2 Cations
paper size	11 x 19.5 cm	9.5 x 14 cm
beaker size	600 mL	400 mL
solvent	40 mL premixed 1:1:1 1-butanol, ethanol, 2 M ammonia	prepare immediately before use - 7 mL of 6 M HCl + 25 mL acetone, thoroughly mix
distance between spots	1.5 cm	2.0 cm
distance of spots from bottom	2.0 cm	2.0 cm

spots

	Brand A[1]		Cations	
1		red		Fe^{3+}
2		blue		Cu^{2+}
3		optional		Co^{2+}
4		green		Mn^{2+}
5		brown		$Fe^{3+}, Cu^{2+}, Co^{2+}, Mn^{2+}$
6		black		unknown
	Brand B[2]			
7		red		
8		blue		
9		optional		
10		green		
11		brown		
12		black		

visualization	dry (hot air blower)	a. dry with hot air blower b. place over cryst. dish of concentrated NH_3 c. dry again
analysis	Report color, spot distance, and R_f value for each pigment in every pen.	After each step (a,b,c), outline each spot and report its color. Determine all R_f values and identify cations in unknown.

---

[1]Flair suggested
[2]Instructor's option

Chemical Capsule

Steel because of strength and low cost is one of the primary construction materials. Steel is an alloy that consists predominantly of iron and up to a few percent of other elements including carbon and manganese. Stainless steel also includes significant percentages of chromium and nickel. Unfortunately iron is susceptible to reaction with oxygen in the presence of water to form rust ($Fe_2O_3 \cdot nH_2O$). The number of waters associated with the iron(III) oxide varies and has been represented by n in the formula. Rusting significantly corrodes and weakens the steel. It is sometimes necessary to take measures to prevent rusting such as galvanizing or painting.

As iron has two common oxidation states, +2 [iron(II) or ferrous] and +3 [iron(III) or ferric], it is not surprising that there is a second oxide of iron, FeO. It is surprising however, that there is a third oxide of iron with the formula Fe_3O_4. This oxide is sometimes called ferrosoferric oxide and occurs in nature as the mineral magnetite. An oxidation number calculation results in a value of +2⅔ for the iron in magnetite. This probably means that two of the three irons in Fe_3O_4 are +3 and the remaining one is +2.

Magnetite is sometimes called lodestone and as indicated by its name is a ferromagnetic material. Most magnets are alloys containing over 50% iron and varying amounts of aluminum, nickel, cobalt, and copper. The element neodymium is one of the best ferromagnetic materials for magnets.

Prelaboratory Exercises - *Experiment 8*

For solutions to the starred problems, see *Appendix A*.

Name_____Date_____Lab Section_____

1. For the first chromatogram (felt tip pens) in this experiment, all of the chromatographed compounds are colored. For the second chromatogram (cations), three of the four cations are distinctly colored but the fourth is not visible. For compounds that do not absorb in the visible region of the spectrum, suggest several ways of locating their spots.

2. Suppose a chromatogram results with two compounds that have the same R_f values. If separation of the two compounds is desired, what variables in the chromatography process could be changed to attempt to achieve separation?

3. Although chromatography is technically a separation technique, chromatography can also be used for identification in some cases. When and how can paper chromatography be used as an identification technique?

4.* Peptides and proteins can be hydrolyzed (broken down) by hydrochloric acid into their component amino acids. The amino acids can then be analyzed by paper or thin layer chromatography. When the synthetic sweetener, aspartame, was hydrolyzed and a chromatogram run along with several standard amino acids, the chromatogram below resulted (the chromatogram below is actually one determined from literature R_f values). Determine the R_f values of the amino acids used and the two spots obtained for aspartame. What are the two amino acids probably linked together in aspartame? [Note: Aspartame is actually a methyl ester of a dipeptide.]

Figure 8-6

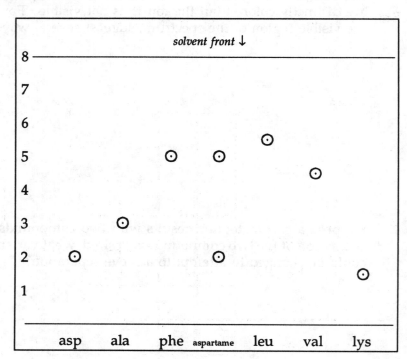

solvent front _____

amino acid	distance	R_f	amino acid	distance	R_f
aspartic acid	_____	_____	leucine	_____	_____
alanine	_____	_____	valine	_____	_____
phenylalanine	_____	_____	lysine	_____	_____
aspartame	_____	_____			
	_____	_____			

aspartame ──HCl─→ CH₃OH + _____ + _____
        methanol

106

Results and Discussion - *Experiment 8*
PAPER CHROMATOGRAPHY

Name_____Date_____Lab Section_____

A. Chromatogram 1 - Ink Pigments

Brand A _____ Brand B_____ Origin to solvent front dist._____

pigments

	spot #		color	dist.	R_f	color	dist.	R_f	color	dist.	R_f	color	dist.	R_f
	1	red												
	2	blue													
A	3	_____													
	4	green													
	5	brown													
	6	black													
	7	red													
	8	blue													
B	9	_____													
	10	green													
	11	brown													
	12	black													

1. Based on color and R_f value, some of the Brand A pigments are used in several different colored felt-tip pens. Give the color and R_f values of the pigments in the red, blue, optional color and green pens that are also apparently used to make the black ink.

black		red			black		blue	
R_f	color	R_f	color		R_f	color	R_f	color
____	_____	____	_____		____	_____	____	_____
____	_____	____	_____		____	_____	____	_____

black		_____(optional)			black		green	
R_f	color	R_f	color		R_f	color	R_f	color
____	_____	____	_____		____	_____	____	_____
____	_____	____	_____		____	_____	____	_____

2. Which pigments, if any, are present in the brown and black Brand A pens and not present in any of the other Brand A pens? Give the color and R_f of any such pigments.

black pen _____brown pen _____

3. Give the color and R_f values of the Brand A pigments that are the same as those used by Brand B.

Brand A pen color	R_f	common color	Brand B pen color	R_f
_____	____	_____	_____	____
_____	____	_____	_____	____
_____	____	_____	_____	____
_____	____	_____	_____	____
_____	____	_____	_____	____
_____	____	_____	_____	____
_____	____	_____	_____	____
_____	____	_____	_____	____
_____	____	_____	_____	____

B. Chromatogram 2. Metal Ions

Unknown # _____   Origin to solvent front _____

soln.	ion	color after drying	color after ammonia	color after 2nd dry	dist. (cm)	R_f	ion
1							
2							
3							
4							
5							
6 unknown	spot 1						
	spot 2						
	spot 3						
	spot 4						

Suggest any ways you can think of to improve any part(s) of this experiment

Experiment 9

CLASSIFICATION OF CHEMICAL REACTIONS

Learning Objectives

Upon completion of this experiment, students should have learned:
1. A scheme for classfying reactions.
2. How to balance chemical equations.
3. The common observations associated with a chemical change.

Text Topics

The classification of chemical reactions, prediction of reaction products, balancing equations (Malone, Chapter 7)

Comments

This experiment is relatively short and can be done in the same laboratory period as another short experiment. For additional problem solving experience with balancing of equations, see *Exercises 15, 16.*

Discussion

The process of classification often assists with the simplification and solution of problems. Classification of diseases by cause; viral, bacterial or fungal, facilitates proper treatment. The sciences are frequently classified into subdisciplines. Chemistry is often subdivided into organic, inorganic, analytical, theoretical and physical branches. Attempts have been made to classify reactions by type. Later in the course, you will learn how to determine if a reaction is an oxidation-reduction reaction. Today, we will run several reactions and attempt to classify them by the nature of the reaction: combination, decomposition, combustion, single replacement or double replacement.

Combination: The reaction of two substances to form one substance.

$$C(s) + O_2(g) = CO_2(g)$$

$$MgO(s) + H_2O(l) = Mg(OH)_2(s)$$

$$SO_3(g) + H_2O(l) = H_2SO_4(l)$$

$$SrCl_2(s) + 6 H_2O(l) = SrCl_2 \cdot 6H_2O(s)$$

Decomposition: The reverse of combination or the breaking down of one substance into two or more substances.

$$H_2CO_3(aq) = H_2O(l) + CO_2(g)$$

$$HgO(s) = Hg(l) + \tfrac{1}{2} O_2(g)$$

$$Ca(HCO_3)_2(aq) = CaCO_3(s) + H_2O(l) + CO_2(g)$$

$$Mg(OH)_2(s) = MgO(s) + H_2O(g)$$

$$SrCl_2 \cdot 6H_2O(s) = SrCl_2(s) + 6 H_2O(g)$$

Combustion: The rapid reaction of a compound with oxygen. Some combustion reactions are also combination reactions and vice versa.

$$2 CH_4O(l) + 3 O_2(g) = 2 CO_2(g) + 4 H_2O(g)$$

$$2 C_8H_{18}(l) + 25 O_2(g) = 16 CO_2(g) + 18 H_2O(g)$$

Single Replacement: The replacement of an element in a compound by another element originally in elemental form.

reaction	observation
$Cu(s) + 2 AgNO_3(aq) = Cu(NO_3)_2(aq) + 2 Ag(s)$	plating of metallic silver on copper and appearance of blue color in solution
$Mg(s) + H_2SO_4(aq) = MgSO_4(aq) + H_2(g)$	gas evolution
$Br_2(aq) + 2 KI(aq) = 2 KBr(aq) + I_2(aq)$	loss of red bromine color and appearance of brown I_2 color

Double Replacement: An exchange of positive and negative ion partners by two compounds (most commonly involving two ionic compounds) in aqueous solution. These reactions usually proceed when at least one of the products is a compound insoluble in water (precipitate), a gas or a compound that decomposes into a gas, or a slightly ionized compound.

reaction	*observation*
$3\ BaCl_2(aq)\ +\ 2\ Na_3PO_4(aq)\ =\ Ba_3(PO_4)_2(s)\ +\ 6\ NaCl(aq)$	white ppt.
$K_2CO_3(aq)\ +\ 2\ HNO_3(aq)\ =\ 2\ KNO_3(aq)\ +\ H_2O(l)\ +\ CO_2(g)$	gas
$H_2SO_4(aq)\ +\ 2\ KOH(aq)\ =\ K_2SO_4(aq)\ +\ 2\ H_2O(g)$	heat

In addition to recording observations about a reaction and classifying the reaction by type, one should write balanced chemical equations for any reactions that occur. The examples above are all balanced.

The first step in writing a balanced equation is to write the correct formulas for reactants and products. Once this is done, subscripts *must not be changed* to balance the equation as this changes the substance. Balance the equation using coefficients (numbers in front of the formula). A coefficient denotes the relative number of moles of the substance whose formula it precedes. Locate the formula with the largest subscript (not subscripts within a polyatomic ion but subscripts that give the number of atoms or ions per formula unit). For the reaction below, 8 is the largest subscript.

$$C_3H_8(g)\ +\ O_2(g)\ \rightarrow\ CO_2(g)\ +\ H_2O(g)$$

Imagine the coefficient 1 in front of the propane (C_3H_8) and balance the hydrogens and then the carbons.

$$1\ C_3H_8(g)\ +\ O_2(g)\ \rightarrow\ 3\ CO_2(g)\ +\ 4\ H_2O(g)$$

Observe that there are now ten oxygen atoms on the right and that there are two oxygen atoms in an oxygen molecule on the left. Divide the number of atoms needed (10) by the number per molecule or formula unit (2) to arrive at the correct coefficient (5).

$$C_3H_8(g)\ +\ 5\ O_2(g)\ =\ 3\ CO_2(g)\ +\ 4\ H_2O(g)$$

Notice that coefficient of 1 is not written in the final equation but is understood. Also notice the very common technique of leaving the O_2 (or H_2) until last if it is present. The coefficient for O_2 affects the amount of one element only whereas the other coefficients change the amounts of at least two elements.

114

For double replacement reactions, start with the ion with the largest subscript. In case of a tie, choose the ion with the largest oxidation number. For the reaction below,

$$BaCl_2(aq) + Na_3PO_4(aq) \rightarrow Ba_3(PO_4)_2(s) + NaCl(aq)$$

Na^+ and Ba^{2+} have subscripts of 3 but Ba^{2+} has the higher oxidation number. Notice that the subscript 4 is part of the phosphate polyatomic ion and is not part of this consideration. Start with the Ba^{2+} and work from one side of the equation to the other. Balance the barium ions on the left side of the equation by inserting a coefficient of 3.

$$3\ BaCl_2(aq) + Na_3PO_4(aq) \rightarrow 1\ Ba_3(PO_4)_2(s) + NaCl(aq)$$

The coefficient of 3 in front of $BaCl_2$ locked in 6 chlorides so a 6 is now needed in front of NaCl on the right.

$$3\ BaCl_2(aq) + Na_3PO_4(aq) \rightarrow 1\ Ba_3(PO_4)_2(s) + 6\ NaCl(aq)$$

The 6 locks in 6 sodiums so the coefficient in front of Na_3PO_4 is the number needed (6) divided by the number per formula unit (3) resulting in a coefficient of 2. Finally check to see if the phosphates are balanced to be sure you haven't made an error.

$$3\ BaCl_2(aq) + 2\ Na_3PO_4(aq) = Ba_3(PO_4)_2(s) + 6\ NaCl(aq)$$

Procedure

Carry out the reactions as instructed. Record all your observations (precipitate, gas evolution, heat evolution, or color change). Use your observations, the nature of the reactants and the examples given in the *Discussion* section to classify the reactions by type. In particular, be alert for combination reactions when two compounds react to form one, decomposition when one compound decomposes into two or more compounds, combustion when a compound reacts with oxygen, single replacement when an element reacts with a compound to give another element and compound and double replacement when two compounds react to give two new compounds (detectable by formation of a precipitate, a gas, or heat evolution). After mixing, touch the exterior of each test tube to check for heat evolution. Write balanced equations for all observed reactions. If a reaction is not detected, write "NAR" for no apparent reaction. Be sure to give each mixture ample time (at least 5 minutes) before concluding there is no apparent reaction (by vision and/or touch). For sample analyses of reactions, see the *Prelaboratory Exercises*.

1. Mix 2 mL of 0.1 M $CaCl_2$ with 2 mL of 0.1 M Na_3PO_4.

2. Add a few drops of water to a test tube containing about 0.5 g $CuSO_4$ (anhydrous). [Caution: Anhydrous copper sulfate (white color) is corrosive. Avoid skin contact but wash with copious quantities of water if contact occurs.]

3. Heat a test tube containing about 0.5 g $Cu(OH)_2$ with a burner.

4. To a test tube containing 3 mL of 6 M HCl, add a 1 cm^2 piece of zinc foil. If a gas evolves, point the test tube away from all people and quickly insert a lighted splint into the mouth of the test tube. Balance the equations for both reactions.

5. Mix 2 mL of 3 M HCl with 2 mL of 1 M Na_2CO_3 [Note that in this case, two types of reactions occur, one right after the other].

6. To a test tube containing 3 mL of 3% H_2O_2 (hydrogen peroxide), add 0.1 g of the catalyst MnO_2. [Note that a catalyst affects the rate of a reaction but is not involved in the overall reaction.]

7. Add 2 mL of a saturated calcium acetate solution (about 35 g/100 mL H_2O) to an evaporating dish. To the dish add 15 mL of ethanol and swirl the contents. Pour off any excess liquid and ignite the remaining contents with a match. For an additional effect, sprinkle some boric acid on the mixture. Although the reaction is actually more complex, assume that the reactants in this reaction are only ethanol (C_2H_5OH) and oxygen.

8. Add 3 g of NH_4Cl and 7 g of $Sr(OH)_2·8H_2O$ to a 125 mL Erlenmeyer flask and swirl vigorously for about 5 minutes. Be sure to record all observations including odor, sounds and touch (flask, not contents) sensations in addition to visual.

9. Mix 2 mL of 3 M H_2SO_4 with 4 mL of 3 M NaOH.

10. Mix 2 mL of 0.1 M $CaCl_2$ with 2 mL of 0.1 M Na_2CO_3.

11. To a test tube containing 3 mL of 0.1 M $CuSO_4$, add a 1 cm^2 piece of zinc foil.

12. To a test tube containing 3 mL of 0.1 M $ZnSO_4$, add a 2 cm long piece of copper wire.

13. Mix 1 mL of 0.1 M $CaCl_2$ and 2 mL of 0.1 M $NaNO_3$.

14. Heat a test tube containing about 2 g of $CuSO_4·5H_2O$ with a burner.

15. Mix 2 mL of 6 M HCl with 4 mL of 3 M NaOH.

Chemical Capsule

On first thought, it would not seem like there would be a significant connection between ozone (O_3), and freons (e.g., $CClF_3$, CCl_2F_2). However, the connection is so important that it threatens to disrupt the delicate ecological balance that exists on earth. Ozone is an extremely active form of oxygen that has both positive and dangerous properties. Because it is a much stronger oxidizing agent than O_2, it reacts much more indiscriminately and will cause damage to many kinds of materials such as human tissue and tires. Consequently, we want the ozone concentration to be minimal in the air we breathe. Unfortunately, cars produce NO which is oxidized in the air to NO_2. The NO_2 is broken down by sunlight to NO and O. The O atoms react with O_2 to give O_3. On days with the appropriate weather, the ozone concentration can increase in smog to dangerous levels. The strong oxidizing power of ozone can also be used. In fact, ozone is commonly used to kill bacteria in water purification plants and has many advantages over chlorine.

While ozone is an undesirable gas in the lower atmosphere, it has a very important and useful role in the upper atmosphere. Ozone absorbs light from the sun in the ultraviolet region that is capable causing considerable damage to living things. For humans, a decrease in ozone permits more skin cancer causing ultraviolet to reach us. For plants, the impact is not completely known but an increase in ultraviolet could significantly affect the life cycle of plankton. As plankton is at the top of the food chain, this kind of an effect would affect all life on earth.

Like ozone, freons also have useful properties. Because they are extremely inert and have ideal properties as refrigerants, freons have been used extensively in all kinds of cooling equipment including air conditioners and refrigerators. Unfortunately, the inertness causes a significant problem. Due to their great stability, freons do not react when released in the atmosphere and after a period of several years diffuse up in the atmosphere to the ozone layer. In the upper atmosphere, freons absorb ultraviolet light that causes their decomposition. The decomposition products initiate a chain reaction that leads to the destruction of ozone.

There is strong scientific evidence that the ozone concentration has decreased on a global scale. In locations such as the Antarctic, ozone decreases annually to the point where an ozone hole is produced. The damage already done will probably lead to an increase of skin cancer but potentially even more dangerous will be the effect on our food chain. Fortunately, most countries have signed an agreement that is phasing out the use of freons and measurements reported in 1996 did give some reason for optimism.

Prelaboratory Exercises - *Experiment 9*

For solutions to the starred problems, see *Appendix A*. For additional problem solving experience with balancing of equations, see *Exercises 15, 16*.

Name_____Date_____Lab Section_____

1. Classify the following reactions according to combination (CA), decomposition (D), combustion (CU), single replacement (SR) or double replacement (DR) and then balance the equations.

reaction	**classification**
a.* __Mg(s) + __ZnCl$_2$(aq) = __MgCl$_2$(aq) + __Zn(s)	_____
b.* __AgNO$_3$(aq) + __CaCl$_2$(aq) = __AgCl(s) + __Ca(NO$_3$)$_2$(aq)	_____
c.* __C$_2$H$_6$(g) + __O$_2$(g) = __CO$_2$(g) + __H$_2$O(g)	_____
d.* __Na$_2$O(s) + __H$_2$O(l) = __NaOH(aq)	_____
e.* __KClO$_3$(s) = __KCl(s) + __O$_2$(g)	_____
f. __Al(s) + __HCl(aq) = __AlCl$_3$(aq) + __H$_2$(g)	_____
g. __C$_3$H$_6$O(g) + __O$_2$(g) = __CO$_2$(g) + __H$_2$O(g)	_____
h. __Fe(s) + __O$_2$(g) = __Fe$_2$O$_3$(s)	_____
i. __Cl$_2$(aq) + __KBr(aq) = __Br$_2$(aq) + __KCl(aq)	_____
j. __Ca(NO$_3$)$_2$(aq) + __K$_3$PO$_4$(aq) = __Ca$_3$(PO$_4$)$_2$(s) + __KNO$_3$(aq)	_____
k. __Ca(HCO$_3$)$_2$(aq) = __CaCO$_3$(s) + __H$_2$O(l) + __CO$_2$(g)	_____
l. __NaOH(aq) + __H$_3$PO$_4$(aq) = __Na$_3$PO$_4$(s) + __H$_2$O(aq)	_____

2. Complete, balance and classify the following reactions:

reaction	classification

a.*　　___$BaCl_2$(aq) + ___Na_2SO_4(aq) =　　　　　　　　　　　　　　　___

b.*　　___Fe(s) + ___$CuCl_2$(aq) =　　　　　　　　　　　　　　　　　___

c.*　　___C_6H_6(l) + ___O_2(g) =　　　　　　　　　　　　　　　　　___

d.*　　___$BaCl_2$(s) + ___H_2O(l) =　　　　　　　　　　　　　　　　___

e.　　___$Pb(NO_3)_2$(aq) + ___K_2CrO_4(aq) =　　　　　　　　　　　___

f.　　___Mg(s) + ___H_2SO_4(aq) =　　　　　　　　　　　　　　　___

g.　　___CaO(s) + ___H_2O(l) =　　　　　　　　　　　　　　　　　___

h.　　___C_2H_4O(l) + ___O_2(g) =　　　　　　　　　　　　　　　___

i.　　___HNO_3(aq) + ___$Ba(OH)_2$(aq) =　　　　　　　　　　　　___

j.　　___HCl(aq) + ___$KHCO_3$(aq)　　　　　　　　　　　　　　　___

Results and Discussion - *Experiment 9*
CLASSIFICATION OF CHEMICAL REACTIONS

Name_____Date_____Lab Section_____

For each system record significant observations (precipitate, gas evolution, heat evolution, color change). For numbers 1-9 classify the potential reaction according to combination (CA), decomposition (D), combustion (CU), single replacement (SR), double replacement (DR), or *NAR* (no apparent reaction) and balance the equations. For numbers 10-15, complete, balance and classify any reactions that occur.

reaction   classification

1. __$CaCl_2$(aq) + __Na_3PO_4(aq) = __$Ca_3(PO_4)_2$(s) + __NaCl(aq)   _____

   observations-

2. $CuSO_4$(s) + __H_2O(l) = $CuSO_4 \cdot$__H_2O(s)   _____

   observations-

3. __$Cu(OH)_2$(s) = __CuO(s) + __H_2O(g)   _____

   observations-

4. __Zn(s) + __HCl(aq) = __$ZnCl_2$(aq) + __H_2(g)   _____

   observations-

   __H_2(g) + __O_2(g) = __H_2O(g)   _____

   observations-

119

reaction	classification

5. __HCl(aq) + __Na$_2$CO$_3$(aq) = __NaCl(aq) + __H$_2$O(l) + __CO$_2$(g) _____

observations-

6. __H$_2$O$_2$(aq) = __H$_2$O(l) + __O$_2$(g) _____

observations-

7. __C$_2$H$_6$O(l) + __O$_2$(g) = __CO$_2$(g) + __H$_2$O(g) _____

observations-

8. __Sr(OH)$_2$·8H$_2$O(s) + __NH$_4$Cl(s) = __SrCl$_2$(aq) + __NH$_3$(aq) + __H$_2$O(l) _____

observations-

9. __H$_2$SO$_4$(aq) + __NaOH(aq) = __Na$_2$SO$_4$(aq) + __H$_2$O(l) _____

observations-

10. __CaCl$_2$(aq) + __Na$_2$CO$_3$(aq) = _____

observations-

<u>reaction</u>   <u>classification</u>

11. __Zn(s) + __CuSO$_4$(aq) =   _____

   observations-

12. __Cu(s) + __ZnSO$_4$(aq) =   _____

   observations-

13. __CaCl$_2$(aq) + __NaNO$_3$(aq) =   _____
   observations-

14. __CuSO$_4$·5H$_2$O(s) =   _____

   observations-

15. __HCl(aq) + __NaOH(aq) =   _____

   observations-

16. Suggest any ways you can think of to improve any part(s) of this experiment.

Experiment 10

EMPIRICAL FORMULA
OF A HYDRATE

Learning Objectives

Upon completion of this experiment, students should have learned:
1. A method of determining the identity of an inorganic salt.
2. To determine the percent water and empirical formula of a hydrate.
3. To use a platinum wire flame tester.

Text Topics

The mole, percent composition, empirical formula, hydrates (Malone, Chapter 8)

Comments

Part C-3 of this experiment should be started first and the remaining parts should be performed when time permits. For additional problem solving experience with the mole, percent composition and empirical formulas, see *Exercises 13, 14.*

Discussion

Many compounds decompose when heated before their melting points are reached. Decomposition of a compound results in the formation of two or more distinct substances. A particular class of compounds that undergo decomposition are the inorganics (often called salts) that contain waters of hydration in their crystal structures. In these compounds called hydrates, the water is weakly bonded to the salt and usually can be driven off with relatively mild heating. The formulas of hydrates are written in a unique fashion. The formula of the anhydrous salt is written first and a dot follows. A coefficient denoting the number of waters per formula unit follows the dot, then the formula, H_2O, is written last. The formula for copper(II) sulfate pentahydrate is written: $CuSO_4 \cdot 5H_2O$. Hydrates are the only compounds that have coefficients in the middle of their formulas.

In this experiment, you will explore some of the properties of hydrates, determine the percent by mass of water in a hydrate (simply by heating it and driving off the water) and the empirical formula (the number of waters in the formula) of a hydrate.

Compounds that absorb water from the air and can be used as desiccants are said to be hygroscopic or deliquescent. In some hydrates, the water of hydration is bonded so weakly that it tends to escape even at room temperature when the compound is exposed to the atmosphere. These compounds are efflorescent. In the first and second parts of this experiment you will study the efflorescence and deliquescence of two compounds and attempt to verify that $CuSO_4 \cdot 5H_2O$ loses water when heated.

In the third part of the experiment, you will be given an unknown hydrate. Using simple qualitative tests, you will determine the identity of the salt. By quantitatively driving off the water, you will be able to calculate the mass percent of water in the hydrate. Knowing the identity of the salt, it will be possible to determine the number of waters of hydration in the formula.

To help you determine the identity of the salt, you will perform a flame test. In the flame test, a platinum wire is inserted into a solution containing the ions of interest and then inserted into a Bunsen burner flame. For some cations, the energy of the flame will cause electrons to be elevated from the ground state or lowest possible energy levels to higher energy orbitals. As this results in an unstable situation or an **excited state**, the electron drops back to the lower energy state and gives off the excess energy often in the form of a characteristic emission of light. The color of the light is useful for the detection of the presence of ions of sodium, barium, calcium, strontium, lithium and potassium. To complete your identification, it may be necessary to attempt one single replacement reaction.

To determine the mass percent of water in the hydrate, the hydrate is weighed, heated and reweighed. The mass percent of water can be calculated from:

$$\frac{\text{weight loss during heating}}{\text{original mass}} \times 100\% = \text{mass \% water in hydrate}$$

Assume that heating of a 3.50 g sample of the hydrate of copper sulfate yields 2.25 g of anhydrous copper sulfate. The mass percent of water in the hydrate would be:

$$\frac{3.50 - 2.25}{3.50} \times 100\% = 35.7\% \text{ [The theoretical value is 36.1\%]}$$

Now the ratio of the number of water molecules per formula unit of copper sulfate can be determined.

$$2.25 \text{ g CuSO}_4 \times \frac{1 \text{ mol}}{159.6 \text{ g CuSO}_4} = 1.41 \times 10^{-2} \text{ mol CuSO}_4$$

$$1.25 \text{ g H}_2\text{O} \times \frac{1 \text{ mol}}{18.02 \text{ g H}_2\text{O}} = 6.94 \times 10^{-2} \text{ mol H}_2\text{O}$$

$$\frac{0.0694 \text{ moles H}_2\text{O}}{0.0141 \text{ moles CuSO}_4} = 4.92 \text{ [This is within experimental error of the actual number of 5]}$$

This indicates there were 5 moles of water per mole of $CuSO_4$.

$$CuSO_4 \cdot 5H_2O(s) \rightarrow CuSO_4(s) + 5 H_2O(g)$$

Procedure

A. Deliquescence and efflorescence. Place a few crystals of sodium sulfate decahydrate on a watch glass. Occasionally observe their appearance for about an hour and write an equation for the observed change. On another watch glass, place a few crystals of potassium acetate. As before, observe their appearance over at least a one hour time period.

B. Copper(II) sulfate pentahydrate. Put 1.0 g of $CuSO_4 \cdot 5H_2O$ into a 13 × 100 mm test tube. Stuff a small wad of fine glass wool [*Caution: Minimize contact of the glass wool with your hands - it causes splinters and itching*] into the test tube so that it holds copper(II) sulfate pentahydrate in place when the tube is tilted downward. Clamp the test tube upside down over a watch glass. Holding the base of the burner, heat the $CuSO_4 \cdot 5H_2O$. The blue color should begin to dissipate while the vapors given off condense to a liquid. If any sign of blackening occurs, reduce the heat. Heat until several drops of liquid have been collected and the residue is white.

Using an eyedropper, test a drop of the liquid collected in the watch glass with a piece of blue cobalt chloride test paper (if the paper is pink, pass it quickly over the burner flame to turn it blue). Test a drop of water on a piece of blue cobalt chloride paper.

Remove the glass wool plug with a wire hook, break up the white residue with a stir rod or wire and pour it into another watch glass [*Caution: avoid letting this powder come into contact with your skin*]. Using a spatula, divide the powder into two piles. Test one pile with a drop of liquid collected in the watch glass. Test the other pile with a drop of water. Write equations for the observed changes.

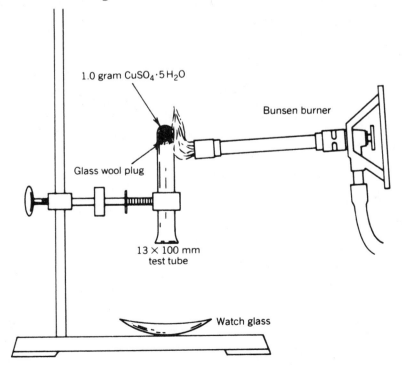

Figure 10-1

C. Analysis and percent water of an unknown hydrate.

1.   Your unknown is a hydrate of strontium chloride, magnesium sulfate or zinc sulfate. Dissolve a small amount (about 0.1 g) in about 5 mL of deionized water in a test tube to use for #1 and #2. Clean a platinum wire by alternately dipping it into 6 M HCl and inserting into the flame until little or no color is observed. Perform flame tests on known solutions of $ZnSO_4$, $MgSO_4$ and $SrCl_2$ and finally the solution of the unknown being sure to clean the wire between each test. This test should enable you to either identify strontium ion or rule it out.

2.   If your unknown is not strontium chloride, drop a short piece of magnesium ribbon into your unknown solution. Allow 30 seconds for a reaction to take place. If the magnesium ribbon tarnishes, your unknown is zinc sulfate.

$$Mg(s) + ZnSO_4(aq) = MgSO_4(aq) + Zn(s)$$

If no reaction other than bubbling occurs ($Mg(s) + MgSO_4(aq)$ = no reaction), the unknown is magnesium sulfate. The bubbling is due to a side reaction of the magnesium with water.

$$Mg(s) + 2 H_2O(l) = Mg(OH)_2(s) + H_2(g)$$

3.   Weigh a clean, dry crucible and cover to at least the nearest 0.01 g. Place about 4 grams (3.5 - 4.5) of your unknown hydrate crystals in the crucible and weigh to at least the nearest 0.01 g. Suspend the crucible in a clay triangle over a Bunsen burner. Heat gently with the top *slightly ajar* for about eight minutes and then vigorously (the crucible should glow a dull orange) for an additional eight minutes. Allow the crucible to cool (about 5 minutes), and weigh it to at least the nearest 0.01 g. Heat the crucible vigorously again for about 5 minutes, cool and weigh again. If the mass difference between the first and second heatings is greater than 0.02 g, perform a third heating, cooling, weighing cycle. Repeat the process until two successive weighings do not differ by more than 0.02 g. Calculate the mass percent of water and the empirical formula of the hydrate.

Figure 10-2

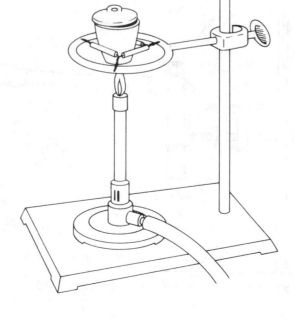

Prelaboratory Exercises - *Experiment 10*

For solutions to the starred problems, see *Appendix A*. For additional problem solving experience with the mole, percent composition and empirical formulas, see *Exercises 13, 14*.

Name_____Date_____Lab Section_____

1. Complete and balance the following reaction:

    $CaCl_2 \cdot 2H_2O(s) \quad —\Delta—>$

2. Determine the following:

    a.* The formula mass of $Cu(NO_3)_2$ _____

    b. The formula mass of $Cu(NO_3)_2 \cdot 3H_2O$ _____

    c.* The number of moles in 2.3 g of ethanol (C_2H_6O) _____

    d. The number of moles in 0.476 g of $NiCl_2 \cdot 6H_2O$ _____

    e.* The mass in grams of 8.7×10^{-4} moles of $CuSO_4 \cdot 5H_2O$ _____

    f. The mass in grams of 4.5×10^{-3} moles of $CrCl_3 \cdot 6H_2O$ _____

    g.* The mass percent of carbon, hydrogen and oxygen in isopropyl alcohol (C_3H_8O) C _____

    H _____

    O _____

    h. The mass percent of oxygen in $Al(NO_3)_3$ _____

    i.* The mass percent of water in $CaCl_2 \cdot 2H_2O$ _____

    j. The mass percent of water in $Co(C_2H_3O_2)_2 \cdot 4H_2O$ _____

3.*   Determine the empirical formula and the molecular formula of a
        compound (Freon 11) that consists of 8.74% carbon, 77.43% chlorine
        and 13.83% fluorine and has a molecular mass of 137 g/mol.   _____

4.   Determine the empirical formula and the molecular formula of a
        compound that consists of 23.54% carbon, 1.98% hydrogen
        and 74.48% fluorine and has a molecular mass of 102 g/mol.   _____

5.   The mass percent of water in a hydrate of $MnCl_2$ is 36.41%. What
        is the empirical formula of the hydrate?   _____

6.*   A 4.00 gram sample of a hydrate of nickel(II) bromide loses
        0.793 grams of water when heated. Determine the mass percent
        of water in the hydrate and the formula of the hydrate.   _____

                                                                            _____

7.   A 2.500 gram sample of a hydrate of calcium sulfate loses
        0.523 grams of water when heated. Determine the mass percent
        of water in the hydrate and the formula of the hydrate.   _____

                                                                            _____

Results and Discussion - *Experiment 10*
EMPIRICAL FORMULA OF A HYDRATE

Name_____Date_____Lab Section_____

A. Deliquescence and efflorescence.

System	Observations	Reaction
1. $Na_2SO_4 \cdot 10H_2O$		
2. $KC_2H_3O_2$		

B. Copper(II) sulfate pentahydrate

System	Observations	Reaction
1. water + $CoCl_2$ paper		
2. condensate + $CoCl_2$ paper		
3. water + $CuSO_4$		
4. condensate + $CuSO_4$		

5. What evidence supports the conclusion that the condensate is water?

C. Analysis and percent water of an unknown hydrate.

1. Unknown #   _____

2. Flame tests

solution	flame color
$SrCl_2$	_____
$ZnSO_4$	_____
$MgSO_4$	_____
unknown	_____

Flame test conclusion _____

3. Mg + unknown → observations _____

   Reaction _____

4. Unknown: $SrCl_2$ or $ZnSO_4$ or $MgSO_4$ _____

5.
   a. Mass of crucible + cover   _____

   b. Mass of crucible + cover + unknown   _____

   c. Mass of crucible + cover + unknown after first heating   _____

   d. Mass of crucible + cover + unknown after second heating   _____

   e. Mass of crucible + cover + unknown after third heating (if necessary)   _____

   f. Mass of crucible + cover + unknown after fourth heating (if necessary)   _____

   g. Mass of original unknown   _____

   h. Mass of water lost by unknown   _____

   i. Mass percent of water in unknown   _____

   j. Mass of unknown salt remaining after heating   _____

   k. Formula mass of anhydrous salt   _____

   l. Moles of anhydrous unknown salt   _____

   m. Formula mass of water   _____

   n. Moles of water lost   _____

   o. Ratio of moles of water to moles of unknown
      (use appropriate number of significant figures)   _____

   p. Formula of unknown hydrate   _____

   q. Unknown identification number   _____

6. Suggest any ways you can think of to improve any part(s) of this experiment.

Experiment 11

STOICHIOMETRY OF A REACTION

Learning Objectives

Upon completion of this experiment, students should have learned:
1. The principles of stoichiometry.
2. The technique of quantitative gravity filtration.
3. The concepts of theoretical and percent yields.

Text Topics

Formula mass, the mole, stoichiometry, experimental yield, percent yield (Malone, Chapter 8)

Comments

While the laboratory manipulations of this experiment should not require more than two hours, the best results are obtained if the product is dried several hours and preferably overnight before the final weighing. It is recommended that students return the day after lab for the final weighing. For additional problem solving experience with moles and stoichiometry, see *Exercises 13, 17.*

Discussion

Analytical chemistry is the branch of chemistry that involves the identification of the substances present in a material (qualitative analysis) and the amounts of each substance present (quantitative analysis). In this experiment, you will perform a quantitative analysis technique that potentially could be used for the determination of the amount of copper in a sample. Using a two-step reaction sequence copper(II) oxide will be synthesized from a weighed amount of copper(II) sulfate pentahydrate. Because the starting material will be a pure compound, $CuSO_4 \cdot 5H_2O(s)$, it will be possible to check the validity of the technique by comparing the amount of copper(II) oxide predicted theoretically to the amount experimentally obtained.

Copper(II) sulfate pentahydrate will be dissolved in water and reacted using a double replacement reaction with sodium hydroxide. The addition of hydroxide ions to a solution containing copper(II) ions results in the precipitation of copper(II) hydroxide.

$$CuSO_4(aq) \ + \ 2\,NaOH(aq) \ = \ Cu(OH)_2(s) \ + \ Na_2SO_4(aq)$$

Subsequent heating *in situ* of the copper(II) hydroxide results in decomposition to copper(II) oxide and water.

$$Cu(OH)_2(s) \ \xrightarrow{\ \ \Delta\ \ } \ CuO(s) \ + \ H_2O(l)$$

The CuO can be quantitatively filtered, dried and weighed. The overall reaction for the sequence is:

$$CuSO_4 \cdot 5H_2O(s) \ + \ 2\,NaOH(aq) \ = \ CuO(s) \ + \ Na_2SO_4(aq) \ + \ 6\,H_2O(l)$$

You will perform the sequence above with an accurately weighed amount of $CuSO_4 \cdot 5H_2O(s)$. From this amount, it is possible to calculate the amount of copper(II) oxide that should be formed (the theoretical yield). By performing the experiment, the experimental yield is obtained and this value is compared with the theoretical yield. The ratio of the experimental to the theoretical (times 100%) is the percent yield. A percent yield close to 100% indicates that the technique could be used as quantitative analysis technique for copper in a sample.

Procedure

Weigh between 1.8 and 2.2 grams of copper(II) sulfate pentahydrate into a 250 mL beaker to at least the nearest 0.01 gram. Add 10 mL of deionized water to the beaker and dissolve the copper salt by swirling. Add 10 mL of 6.0 M NaOH to the solution with stirring. Place a watch glass over the beaker and heat the mixture to the boiling point. Try to avoid spattering especially onto the watch glass. If spattering occurs, use a wash bottle to wash all the solid back down into the solution. Heat until all of the blue solid has been decomposed to copper(II) oxide and water (a few minutes). Allow the mixture to cool before filtering.

Fold a 12.5 cm diameter piece of Whatman #1 filter paper (see *Figure 11-1*). Tear off a small outside corner of the paper to improve filtering. Weigh and record the mass of the paper. Open up the "pocket" of the filter paper that doesn't have the corner torn off so that the filter paper is in the form of a cone and place it in the funnel. Using your wash bottle, wet the paper so that it adheres tightly to the sides of the funnel. Support the funnel with a ringstand and ring over a beaker.

Transfer the previously heated mixture to the filter. Be sure not to overload the filter (the liquid should not rise to less than 0.5 cm from the top of the paper). Be patient and add small portions. After all liquid has been transferred to the funnel, use a wash bottle to direct spurts of water at the remaining solid in the beaker and transfer this mixture to the funnel. Continue this procedure until virtually all the solid (precipitate) has been transferred. When the liquid has drained from the filter paper, "wash" the precipitate by directing a stream of water from a wash bottle into it. The stream should be forceful enough to agitate the precipitate but not so forceful that it spatters it. This technique removes any soluble sodium sulfate or sodium hydroxide that may be trapped in the copper oxide precipitate. When the wash water has drained, wash the precipitate once more with water and allow it to filter through until dripping ceases. [Note: It is possible to speed up the next step (drying) by now washing the sample with acetone. However, acetone is a fire hazard and drying takes at least one hour even with an acetone washing. For these reasons we strongly prefer to avoid the acetone washing and to assure complete drying by leaving it in the oven overnight.]

Discard the filtrate which should be colorless and not contain any precipitate. Carefully lift out the paper and unfold it on a watch glass. Make sure that the paper and watch glass are properly labeled and place the watch glass in a 105°C oven for a minimum of 3 hours (preferably overnight). Remove and weigh. If a shorter heating time is used, it is advisable to perform another heating and weighing cycle. Calculate the percent yield.

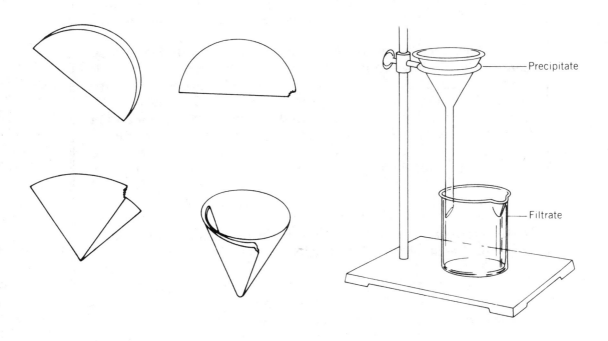

Figure 11-1

In Las Vegas, you might see:
$HClO_2(aq)$ $HClO_2(aq)$ $HClO_2(aq)$ $HClO_2(aq)$ $HClO_2(aq)$ $HClO_2(aq)$ $HClO_2(aq)$ $HClO_2(aq)$

Chemical Capsule

To fluoridate or not to fluoridate a water supply is a question that has puzzled and even agitated many communities. There is not much doubt that fluorides afford some protection against tooth decay. Teeth are protected by a 2 mm layer of a mineral called hydroxyapatite, $Ca_5(PO_4)_3OH$. Hydroxyapatite slowly dissolves or demineralizes according to the following equation:

$$Ca_5(PO_4)_3OH(s) \; = \; 5\,Ca^{2+} \; + \; 3\,PO_4^{3-} \; + \; OH^-$$

The body fights this decay by running the reverse of the above reaction or mineralization and a balance is hopefully maintained.

However, some foods, especially those with high sugar content, break down in the mouth to produce acids such as acetic acid and lactic acid. The acids react with the base produced in the demineralization reaction to produce water. This causes the demineralization reaction to shift to the right resulting in a loss of protective enamel. Without the enamel, tooth decay can begin.

With fluoride present, the following reaction can take place:

$$5\,Ca^{2+} \; + \; 3\,PO_4^{3-} \; + \; F^- \; = \; Ca_5(PO_4)_3F(s)$$

As the product, fluorapatite contains fluoride rather than hydroxide, it is more resistant to acid and helps to protect the teeth.

As with all other chemicals, too much is not good either. While fluoride is safe at the concentrations needed to provide protection, higher levels cause mottled teeth and are toxic. Fluoride is present as SnF_2 or NaF in many toothpastes. Dentists give fluoride treatments for teeth and some municipalities fluoridate water supplies. The total exposure to fluoride from these sources should be kept at a useful but safe level.

A chlorous line!

Prelaboratory Exercises - *Experiment 11*

For solutions to the starred problems, see *Appendix A*. For additional problems on the mole and stoichiometry, see *Exercises 13, 17*.

Name_____Date_____Lab Section_____

1. Determine the following:

    a. The formula mass of $Mg(NO_3)_2$ _____

    b.* The formula mass of $Cu(NO_3)_2 \cdot 6H_2O$ _____

    c. The number of moles in 5.4 g of water   _____

    d.* The number of moles in 3.6×10^{-2} g of $K_2C_2O_4 \cdot H_2O$ _____

    e. The mass in grams of 5.1×10^{-2} moles of $ZnSO_4 \cdot 7H_2O$ _____

    f.* The mass in grams of 4.5 moles of $Fe(NO_3)_3 \cdot 9H_2O$ _____

2.* Aluminum reacts with hydrochloric acid to give aluminum chloride and hydrogen gas according to the following equation:

    $$2\ Al(s)\ +\ 6\ HCl(aq)\ =\ 2\ AlCl_3(aq)\ +\ 3\ H_2(g)$$

    If 2.7 grams of aluminum is reacted with an excess of hydrochloric acid, what is the theoretical yield in grams of hydrogen gas?

135

3. Aspirin (acetylsalicylic acid) is synthesized from salicylic acid and acetic anhydride according to the following equation:

$$C_7H_6O_3 \quad + \quad C_4H_6O_3 \quad = \quad C_9H_8O_4 \quad + \quad C_2H_4O_2$$

    salicylic acid   acetic anhydride   aspirin   acetic acid

Reactions involving organic chemicals often do not give stoichiometric amounts of products because of competing reactions and incomplete reaction. Thus the experimental yield is seldom above 90% of the theoretical amount and is usually lower than 90%.

a.* When 69 grams of salicylic acid was reacted with an excess of acetic anhydride, 72 grams of aspirin was obtained. Calculate the theoretical and percent yields for aspirin in this reaction.

_____

_____

b. If 25 g of acetic anhydride had been used in the reaction above, which reagent would have been limiting and what would the theoretical yield of aspirin have been?

_____

_____

4. a. Reaction of 1.5 g calcium chloride with 3.4 g of silver nitrate gives 2.7 g of silver chloride. Write the balanced formula equation and net ionic equation for the reaction, determine the limiting reagent and the theoretical and percent yields of silver chloride.

b. If an unknown amount of silver nitrate is reacted with an excess of calcium chloride and 1.35 g of silver chloride is obtained, how many grams of silver nitrate reacted?

c. What assumption did you make to calculate the answer in 4-b?

5. Give conditions that must be satisfied for the precipitation method of quantitative analysis to be applicable.

Results and Discussion - *Experiment 11*
STOICHIOMETRY OF A REACTION

Name_____Date_____Lab Section_____

1. Write a balanced net ionic equation for the reaction between aqueous copper(II) sulfate and sodium hydroxide.

2. Calculation of the theoretical yield and percent yield of CuO.

   a. Mass of beaker + $CuSO_4 \cdot 5H_2O$ _____

   b. Mass of empty beaker _____

   c. Mass of $CuSO_4 \cdot 5H_2O$ _____

   d. Formula mass of $CuSO_4 \cdot 5H_2O$ _____

   e. Moles of $CuSO_4 \cdot 5H_2O$ _____

   f. Theoretical number of moles of CuO _____

   g. Formula mass of CuO _____

   h. Theoretical yield in grams of CuO _____

   i. Mass of filter paper + CuO after drying _____

   j. Mass of filter paper _____

   k. Experimental mass of CuO _____

   l. Percent yield of CuO _____

   m. Percent deviation between experimental and theoretical yields _____

   n. Show the series of unit conversions that one could use to calculate the theoretical yield in grams of CuO from the grams of $CuSO_4 \cdot 5H_2O$.

3. Why must the CuO precipitate in the funnel be thoroughly washed with water?

4. Could this procedure be used to determine the mass percent of copper in a sample that also contains magnesium or zinc ions? Explain your answer.

5. What would the consequences have been in this experiment if you had used 10 mL of 0.1 M NaOH instead of 10 mL of 6.0 M NaOH?

7. Suggest any ways you can think of to improve any part(s) of this experiment.

Experiment 12

ENTHALPIES IN PHYSICAL AND CHEMICAL CHANGES

Learning Objectives

Upon completion of this experiment, students should have learned:
1. The meaning of "enthalpy".
2. A method for measuring the heat of fusion.
3. The concepts of lattice energy and heat of hydration.
4. The meaning of "limiting reagent".
5. The net ionic equation for strong acid - strong base reactions.

Text Topics

Limiting reagents, enthalpies of fusion, dissolution, reaction (Malone, Chapter 8)

Comments

For additional problem solving experience involving limiting reagents and net ionic equations, see *Exercises 16, 17*.

Discussion

Energy is either evolved or consumed in both physical and chemical changes. The amount of energy required to melt a solid is called the heat of fusion. The amount of energy required to boil a liquid is its heat of vaporization. The heat involved in many changes is also called the enthalpy. Both melting and boiling require energy input and are endothermic. The reverse processes, condensation and freezing evolve energy and are exothermic.

When an ionic compound dissolves in water, energy is required to break up the crystal lattice but energy is released by the formation of bonds between ions and the water (energy of hydration). When the energy of hydration is greater than the crystal lattice energy, the dissolving process will be exothermic and causes a temperature increase. When the crystal lattice energy exceeds the energy of hydration, the dissolving process is endothermic and the temperature of the system will drop.

In chemical reactions, the breaking of bonds requires energy while formation of new bonds gives off energy. The balance determines if a reaction is exothermic or endothermic.

Procedure

A. Enthalpy of fusion of ice. To determine the approximate heat of fusion of ice, you will add some ice to water and measure the temperature change of the water (Δt_d) and of the melted ice (Δt_i). As energy is conserved within the system (neglecting energy transfer involving the styrofoam cup), the sum of the energy changes must equal zero. When the ice is added to the water, the water will lose an amount of energy (ΔE_w) equal to the product of its mass (m_w), its temperature drop (Δt_d), and its specific heat (C_w).

$$\Delta E_w = m_w \times \Delta t_d \times C_w$$

The specific heat of water is the amount of energy required to heat one gram of water one degree Celsius and is equal to 1.00 calorie/g-K or 4.18 joule/g-K. The energy lost by the water originally in the cup will melt the ice and raise the melt-water to the final temperature. The value of this energy is the mass of the water (m_w) times the temperature drop of the water (Δt_d) times the specific heat of the water (C_w). The energy required to melt the ice is the heat of fusion of ice (ΔH_f) times the mass of the ice (m_i). The energy required to raise the temperature of the melted ice to its final temperature is the product of the mass of the melted ice (m_i) times the temperature increase (Δt_{in}) times the specific heat of water (C_w). Summing the energy changes and setting them equal to zero results in:

$$\underset{\text{energy lost by water}}{(m_w \times \Delta t_d \times C_w)} + \underset{\substack{\text{energy used} \\ \text{to melt ice}}}{(\Delta H_f \times m_i)} + \underset{\substack{\text{energy used to raise} \\ \text{temperature of melted ice}}}{(m_i \times \Delta t_{in} \times C_w)} = 0$$

Solving for ΔH_f results in:

$$\Delta H_f = \frac{-(m_w \times \Delta t_d \times C_w) - (m_i \times \Delta t_{in} \times C_w)}{m_i}$$

The *Results and Discussion* section on the heat of fusion will take you through these calculations.

Transfer 100 mL of deionized water into a preweighed styrofoam cup. Weigh the cup and water to at least the nearest 0.01 gram. Measure and record the temperature of the water to the nearest 0.1°C. Put crushed ice in a 150 mL beaker to the 40 mL graduation (alternatively an ice cube with a mass between 10 and 20 grams can be used.) Empty the ice onto a dry paper towel, pat it dry with another paper towel, and quickly transfer the ice to the water in the styrofoam cup. Stir the mixture with your thermometer. When all the ice has melted, record the water temperature to the nearest 0.1°C and weigh the combination [ice melt + water + cup] to at least the nearest 0.01 gram.

B. Enthalpy of solution. The temperature change upon dissolving two different ionic compounds will be measured to determine the relative magnitudes of the crystal lattice energy and the heat of hydration.

1. Place 3 grams of lithium chloride in a test tube and add 15 mL of water. Stopper the tube and shake it vigorously. Measure the temperature.

2. Into a second test tube, place 3 grams of ammonium chloride and 15 mL of water. Stopper the tube and shake it vigorously. Measure the temperature.

      C. Enthalpy of reaction. The endothermicity or exothermicity of several acid - base reactions will be determined by measuring the temperature change upon mixing. As a result of the series of mixtures, you will be able to evaluate the effects of the nature and concentrations of the acids and bases on the enthalpies changes of the reactions.

      Number nine test tubes *1* through *7*. Put a thermometer into test tube #*1*, add 6.0 mL of the first reagent and measure the temperature. Add 6.0 mL of the second reagent, stir carefully with the thermometer and record the maximum temperature. Repeat this procedure with test tubes 2 - 7.

test tube	first reagent (6 mL)	second reagent (6 mL)
1	3 M HCl	3 M NaOH
2	3 M NaOH	3 M HCl
3	3 M HCl	3 M KOH
4	3 M HNO_3	3 M NaOH
5	1.5 M H_2SO_4	3 M NaOH
6	1.5 M HCl	3 M NaOH
7	6 M HCl	6 M NaOH

Chemical Capsule

Among the top twelve industrial chemicals produced in the United States each year are seven that are important because of their acid or base properties. The acids are sulfuric acid (H_2SO_4), phosphoric acid (H_3PO_4) and nitric acid (HNO_3) and the bases are lime (CaO or slaked lime - $Ca(OH)_2$), sodium hydroxide (NaOH) and sodium carbonate (Na_2CO_3). Part of the need for acids and bases results from the observation that hydrogen ions or hydroxide ions function as general catalysts for many reactions. Other reactions consume hydrogen or hydroxide ions. In addition, the acids and bases are important in many other processes including metallurgy, petroleum refining and the synthesis of fertilizers, plastics, detergents, glass, drugs, dyes, paint, paper and explosives.

The other five chemicals produced out of the top twelve are nitrogen, oxygen, chlorine and two organics, ethylene and propylene. Among many applications, nitrogen is used for the synthesis of ammonia which in turn is used for the synthesis of nitric acid. Oxygen is used predominantly in the steel industry but is also used in chemical manufacture, sewage treatment and rocket fuel. Chlorine is used principally in water treatment, bleaching and chemical manufacture. Ethylene (C_2H_4) and propylene (C_3H_6) are used in polymer synthesis and in the preparation of many other organic chemicals.

Of the remaining 38 chemicals in the top 50, over half are organic. Many of the organics such as acetone (solvent) are themselves of use. Others are used for the synthesis of additional organic chemicals such as polymers and medicines. One recent organic entry into the top 50 was methyl *tert*-butyl ether [MTBE - $(CH_3)_3COCH_3$]. Because of its high octane rating and much lower threat to the environment, MTBE has replaced the use of lead compounds in gasoline.

Included among the remaining inorganic chemicals in the top 50 are ammonium nitrate (fertilizers, explosives), carbon dioxide (refrigerant, beverage carbonation), hydrochloric acid (chemical manufacturing) and a topic of an earlier *Chemical Capsule*, titanium dioxide (paint pigment).

Prelaboratory Exercises - *Experiment 12*

For solutions to the starred problems, see *Appendix A*. For additional problem solving experience involving limiting reagents and net ionic equations, see *Exercises 16, 17*.

Name_____Date_____Lab Section_____

1. Processes that evolve heat are said to be exothermic and by convention have negative enthalpy values (ΔH is $-$). Processes that require or consume heat energy are called endothermic and have positive enthalpy values (ΔH is $+$). Give the sign ($+$ or $-$) for each of the following processes.

    a.* Boiling of water   _____

    b.* Burning of wax vapor   _____

    c.   Freezing of water   _____

    d.   Reaction of an acid with a base   _____

    e.   Reaction of hydrogen with oxygen to give water   _____

    f.   Reusable hand warmers often contain a supersaturated solution
         of sodium acetate. Jarring the system causes crystallization of
         the sodium acetate.   _____

    g.   Evaporation of rubbing alcohol from your skin   _____

2.* Ethylene glycol (antifreeze) melts at -11.5°C and the specific heat of its liquid is 2.3 J/g-K. A 15 gram solid chunk (at -11.5°C) of ethylene glycol was put into 1.00×10^2 grams of liquid ethylene glycol (original temperature = 18.0°C). The final temperature after the solid melted was 4.0°C. Calculate the heat (enthalpy) of fusion of ethylene glycol in joules/g.

                                                                   _____

3.   A 55.0 g piece of iron at 100.0°C is immersed in 20.0 g of water originally at 22.0°C. The resulting temperature of the mixture is 39.8°C. Assuming the calorimetry constant is zero, calculate the specific heat of iron in cal/g-deg and joules/g-deg.

                                                                   _____

                                                                   _____

4. Write formula and net ionic equations for the following reactions:

    a.* nitric acid + potassium hydroxide

    b.* hydrochloric acid + strontium hydroxide

    c. acetic acid + potassium hydroxide

    d. sulfuric acid + barium hydroxide

5.* If equal volumes of 1 M NaOH and 1 M H_2SO_4 are mixed, which is the limiting reagent? Explain your answer.

6. Chemical explosives such as dynamite usually undergo reactions that have three common characteristics: the reactions have large negative ΔH values, very fast rates and result in the conversion of relatively small volumes of liquids or solids to huge volumes of gas.

    a.* Name a few common explosives. _____ _____ _____

    b. What common factor do the explosives you named have?

    c. Draw the Lewis structure of N_2.

    d. Based on the Lewis structure of nitrogen, suggest a contributing reason for the enthalpy values of explosive reactions.

Results and Discussion - *Experiment 12*
ENTHALPIES OF PHYSICAL AND CHEMICAL CHANGES

Name_____Date_____Lab Section_____

A. Enthalpy of fusion of ice.

1. Mass of water + cup   _____

2. Mass of cup   _____

3. Mass of water (m_w)   _____

4. Temperature of water before addition of ice (t_o)   _____

5. Temperature of water + ice melt (t_f)   _____

6. Temperature change of water ($\Delta t_d = t_f - t_o$)   _____

7. Temperature change of ice melt ($\Delta t_i = t_f - 0 = t_f$)   _____

8. Mass of water + cup + ice melt   _____

9. Mass of ice (m_i)   _____

$$\underset{\substack{m_w \quad \Delta t_d \quad C_w \\ \textit{energy lost by water}}}{\underline{} \times \underline{} \times 4.18\ \text{J/g-K}} + \underset{\substack{m_i \\ \textit{energy used to} \\ \textit{melt ice}}}{\underline{} \times \Delta H_f} + \underset{\substack{m_i \quad \Delta t_i \quad C_w \\ \textit{energy used to raise} \\ \textit{temperature of melted ice}}}{\underline{} \times \underline{} \times 4.18\ \text{J/g-K}} = 0$$

$$\underline{} + \underline{} \times \Delta H_f + \underline{} = 0$$

$$\underline{} \times \Delta H_f + \underline{} = 0$$

$$\Delta H_f = \underline{}$$

11. The literature value for ΔH_f is 334 J/g. Calculate your percent error.   _____

12. Calculate the molar heat of fusion of water (J/mol)   _____

B. Enthalpy of dissolution.

1. Temperature of 3 grams of lithium chloride added to 15 mL of water _____

2. Temperature of 3 grams of ammonium chloride added to
   15 mL of water _____

3. Which is greater for ammonium chloride, the lattice energy or the
   hydration energy? Explain your answer. _____

C. Enthalpy of reaction.

test tube	1st reagent (6 mL)	temperature (°C)	2nd reagent (6 mL)	temperature (°C)	temp. difference (°C)
1	3 M HCl	_____	3 M NaOH	_____	_____
2	3 M NaOH	_____	3 M HCl	_____	_____
3	3 M HCl	_____	3 M KOH	_____	_____
4	3 M HNO_3	_____	3 M NaOH	_____	_____
5	1.5 M H_2SO_4	_____	3 M NaOH	_____	_____
6	1.5 M HCl	_____	3 M NaOH	_____	_____
7	6 M HCl	_____	6 M NaOH	_____	_____

1. Write net ionic equations for the reactions in tubes 1 - 5.

   #1

   #2

   #3

   #4

   #5

2. Comparing numbers *1* and *2*, does the order of mixing have a
   significant effect on the temperature change? Explain your answer.   _____

3. Comparing numbers *1* and *3*, does the cation of a strong base have
   a significant effect on the temperature change? If not, why not?   _____

4. Comparing numbers *1, 4* and *5,* does the nature of a strong acid have a significant effect on the temperature change? If not, why not? _____

5. Comparing numbers *1* and *6,* was there a significant difference in the temperature change? If so, why? _____

6. Comparing numbers *1* and *7,* was there a significant difference in the temperature change? If so, why? _____

7. Suggest any ways you can think of to improve any part(s) of this experiment.

Experiment **13**

CHEMICAL PROPERTIES OF OXYGEN AND HYDROGEN

Learning Objectives

Upon completion of this experiment, students should have learned:
1. Some techniques of gas collection.
2. Some of the chemical and physical properties of oxygen and hydrogen.

Text Topics

Physical and chemical properties of gases, chemical reactions, stoichiometry (Malone, Chapters 3, 7, 8, 9)

Comments

The hydrogen generation portion of this experiment was derived from an experiment written by David Ehrenkranz and John J. Mauch in *Chemistry in Microscale*, Kendall/Hunt Pub. Co., Dubuque (1990).

Discussion

Oxygen and hydrogen are two of the most important elements in the universe. Hydrogen atoms compose 90% of the atoms of the universe and 75% of the mass. The fusion of hydrogen in the sun is the ultimate energy source for all of life's processes. Constituting 85.5% of the oceans by mass and 47% of the solid crust of the earth, oxygen is the most abundant element in the crust. Oxygen is a constituent of all living tissues and almost all plants and animals require oxygen in the free or combined state to maintain life.

The first known reported preparation of oxygen was by Robert Boyle in 1678 when he decomposed red lead oxide with heat produced using a burning glass. Oxygen, however, was not distinguished from other gases until 1773 when C. W. Scheele in Sweden differentiated it from other gases by its chemical properties. The first published report that oxygen was a distinct type of gas was in 1774 by Joseph Priestley, an Englishman. Priestly showed that the gas he was preparing by decomposing mercuric oxide with heat, was the same gas that was necessary for the respiration of animals. Priestly also showed that green plants in the presence of sunlight, prepare oxygen from carbon dioxide. Because Priestley published his findings before Scheele, Priestley is usually credited with the discovery of oxygen.

Oxygen can be prepared by the decomposition of nitrates (Scheele's method), potassium chlorate, several heavy metal oxides, hydrogen peroxide, and several other compounds. Commercially, it is prepared by distilling it off from liquified air. In this experiment oxygen will be prepared from a reaction of sodium hypochlorite and hydrogen peroxide and on a smaller scale, by a catalyzed decomposition of hydrogen peroxide.

Hydrogen gas was probably first produced by the alchemists. In the sixteenth century, Paracelsus, sometimes referred to as the "father of pharmacy", stated that when an acid acts on iron: "An air arises which bursts forth like the wind." In 1650, T. Turquet de Mayerne reported that hydrogen was inflammable. Robert Boyle prepared hydrogen in 1671 in order to show that it obeyed the same physical laws as other gases. In 1678, Isaac Newton, the discoverer of calculus, the universal law of gravitation, and the inventor of the reflecting telescope, suggested in a letter to Boyle that hydrogen contained particles of iron (until this time hydrogen had been prepared almost exclusively by the action of acids on iron). Newton's hypothesis was disproved in 1766 by Henry Cavendish when he showed that the gases produced by the action of acids on iron, zinc and tin were identical. Before this time, writers generally didn't distinguish between hydrocarbon gases, carbon monoxide, hydrogen sulfide and hydrogen. All were termed 'inflammable air". Cavendish showed that hydrogen had a different specific gravity than the other gases known as inflammable air. Because he demonstrated these facts, Cavendish is usually credited with the discovery of hydrogen.

Hydrogen can be prepared by the action of acids on a number of metals, the action of bases on several active metals, the action of very active metals on water and by electrolysis of water. In today's experiment, hydrogen will be generated by reacting zinc with hydrochloric acid.

Oxygen is one of the most reactive elements and combines with a majority of the other elements. A reaction of an element or compound with oxygen is often called combustion or oxidation (oxidation doesn't have to mean combination with oxygen and more generally means a loss of electrons or an increase in oxidation number). In this experiment you prepare and collect oxygen and study several combustion reactions including oxidation of wax, wood, carbon, iron (would you expect a metal to burn?) and hydrogen. Because of the potentially explosive nature of the combustion of hydrogen, the reaction of hydrogen with oxygen will be run on a small scale in Beral pipets. The hydrogen - oxygen reaction will be performed with varying ratios of hydrogen to oxygen and the optimal mixture will be qualitatively compared to the ratio predicted by the balanced equation.

Procedure

A. The candle flame. Light a candle and study the flame. Record all your observations and questions about the flame. Try to include observations on states of matter and physical and chemical properties and changes. Be very careful to distinguish observations from explanations. What do you think is actually burning? Write your answer down before you read or experiment further. You will not lose points for an incorrect answer. With a match, try burning solid wax, liquid wax and a waxless wick. Put a beaker over the burning candle but not all the way down and carefully observe the inside of the beaker. Blow the candle out and immediately put a glass stirring rod next to the extinguished wick. Inspect the rod. Record observations from these tests. Light the candle again and immediately after blowing out the flame put a lighted match above the wick (do not touch the wick with the match) where the flame used to be.

B. Preparation and properties of oxygen. [Note: this experiment may be performed by the students or given as a demonstration by the instructor.]

<u>Step 1.</u> Set up the apparatus shown in *Figure 13-1*. When constructing the apparatus, use glycerin to lubricate the rubber stopper before inserting glass tubing. Hold the tubing with a towel very close to the stopper as you insert it. Transfer 200 mL of 5.25% chlorine bleach into the 500 mL Florence flask. Add 3% hydrogen peroxide to partially fill the funnel.

<u>Step 2.</u> Fill a dishpan about ⅓ full of water. Fill a gas collecting bottle over the brim with water and slide a glass cover across the mouth of the bottle. Holding the plate tightly over the mouth of the bottle, tip the bottle upside down, immerse it in the dishpan and remove the glass plate. Repeat this process with five more bottles. Try to minimize the amount of air left in the bottle.

<u>Step 3.</u> Insert the collecting tube into an inverted water filled bottle and carefully use the pinchclamp to allow the 3% hydrogen peroxide to slowly enter the Florence flask. The following reaction will occur to generate oxygen:

$$NaClO_{(aq)} \quad + \quad H_2O_{2(aq)} \quad = \quad H_2O_{(l)} \quad + \quad NaCl_{(aq)} \quad + \quad O_{2(g)}$$

Fill the first bottle with the evolving gas and slip a glass cover plate over the mouth the bottle while it is still under water. This will prevent the introduction of air into the bottle. Try to minimize the amount of water left in the bottle. Set the bottle upright with the glass plate still in place over its mouth. Repeat this process for the next four bottles being sure that you number the bottles in order of collection from 1 to 5. It will occasionally be necessary to refill the funnel with the hydrogen peroxide solution and a total of about 100 mL will be required. When you fill the sixth bottle with oxygen, leave a few milliliters of water in the bottle when you withdraw the collecting tube and insert the cover plate.

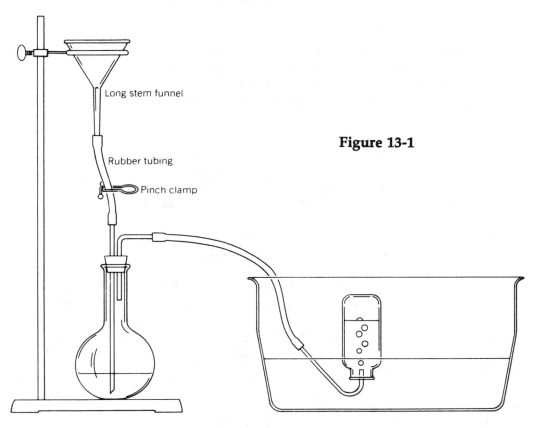

Long stem funnel

Rubber tubing

Pinch clamp

Figure 13-1

Step 4. Place a short candle on a glass plate and light it. After about 1 minute, lower the *first* bottle of oxygen collected over the candle so that the mouth of the bottle sets on the glass plate and measure the amount of time it takes for the candle to burn out. Repeat this procedure, this time using the third bottle of oxygen collected. Repeat this procedure, this time using a bottle of air (a bottle of air is obtained by filling any previously used bottle with water and then dumping out the water).

Step 5. Ignite a wooden splint; blow it out and while it is still glowing, insert it into the second bottle of oxygen collected. Repeat this process with a bottle of air.

Step 6. Keeping the cover plate in place over the mouth of the fourth bottle of oxygen, place the bottle of oxygen upside down over a bottle of air (*Figure 13-2*). Now slip the cover plate from between the two bottles such that the mouth of the bottle of oxygen is directly over the mouth of the air bottle.

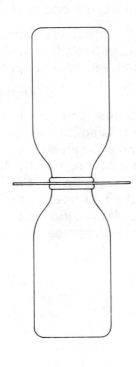

After about eight minutes, slip two cover plates between the two bottles. Place a candle on a glass plate, light it, and after 1 minute with the bottom bottle remaining covered, test the length of time the candle remains burning in the bottle that had been on top. Now test the length of time the candle remains burning in the bottle that had been on the bottom.

Step 7. Hold a marble-sized piece of charcoal with your tongs, ignite it, and thrust it into the fifth bottle of oxygen. Observe the flame.

Step 8. Hold a piece of steel wool with your tongs, pass it through the hottest part of the burner flame and thrust it immediately while it is still glowing into the sixth (the one with a few mL of water) oxygen bottle. Observe the flame.

Figure 13-2

C. **Preparation and properties of hydrogen.** (Because of the potentially explosive nature of hydrogen, these experiments will be conducted on a small scale.)

Step 1. Prepare 2 test tube gas generators (see *Figure 13-3*). One will be used in *steps 1 - 8 and 12, 13* as a hydrogen generator. The second will be used in *steps 9 - 13* as an oxygen generator.

a. Cut off the stem of a graduated Beral pipet (e.g., Flinn Scientific # AP 1516) about 1 cm from the bulb. The bulb will be henceforth referred to as the "collection bulb".

b. Cut the tip of the pipet stem at a point just before the diameter begins to get smaller and insert it (the tip) into a # 2 one hole rubber stopper as shown. Place the stopper in a 20 × 150 mm test tube.

Step 2. Add about 200 mL of water to a 400 mL beaker. This will be used as a temperature regulator during the generation of hydrogen and oxygen.

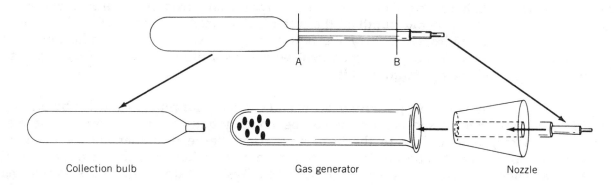

Figure 13-3

Collection bulb   Gas generator   Nozzle

Step 3. Fill the collection bulb with water by holding it (opening upward) under water in a dish pan, squeezing the air out and allowing water to enter.

Step 4. Add about 8 grams of 20 mesh granular zinc to one of the gas generator tubes.

Step 5. Light a candle.

Step 6. Add 25 mL of 2 M hydrochloric acid to the test tube containing the zinc and stopper it with the # 2 stopper device. Put the tube in the 400 mL beaker of water and wait 10 seconds before beginning the next part (see *Figure 13-4*). [Note: If gas generation slows during the experiment, replace the hydrochloric acid but do not replace the zinc]

Figure 13-4

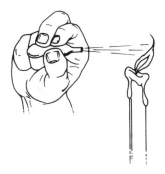

Step 7. Put the collection bulb over the pipet tip (nozzle) and collect hydrogen until all the water has been displaced from the bulb.

Step 8. Remove the bulb from the hydrogen generator and squirt the hydrogen you collected into the candle flame (*Figure 13-5*). Repeat the hydrogen collecting and squirting once or twice and report your observations.

Figure 13-5

Step 9. Add about 7 grams of manganese metal to the second (oxygen) generator tube, label it and fill another collection bulb as in *step 3*.

Step 10. Add 25 mL of 3% hydrogen peroxide to the oxygen generating tube and put this tube in the beaker of water next to the hydrogen generator tube. [Note: If gas generation slows during the experiment, replace the 3% hydrogen peroxide but do not replace the manganese metal.

Step 11. Collect a bulb full of oxygen as in *step 7* and squirt the oxygen into the candle flame. Repeat this process a couple of times and report your observations.

Step 12. Now collect half a bulb full of hydrogen, remove the bulb from the hydrogen generator and put it on the oxygen generator until all the water is displaced from the bulb. This should give a 50/50 mixture of hydrogen and oxygen. Squirt the gas mixture into the candle flame and report your observations. Repeat the experiment with several different ratios of hydrogen to oxygen and try to find the optimum mixture.

Step 13. Refill the pipet with the optimum mixture and convert your pipet to a rocket by placing it on the rocket launcher. A rocket launcher can be made by putting a nail through a piece of cardboard, attaching a long wire to the head of the nail (leave the other end of the wire unattached and out of reach) and putting the "collection bulb rocket" over the point of the nail (*Figure 13-6*). The rocket can be launched by bringing a Tesla coil up to the rocket so that the spark jumps to the nail. [*Caution: Because of potential hazards of electric shock from the coil, the instructor should launch all rockets.*] Have a contest with the other students in the class to determine who can make the farthest flying rocket. Try varying the takeoff angle (nail angle) and the nail diameter (the larger the diameter without creating friction, the better in our experience). **Also try leaving some water in the rocket.**

Step 14. When you are finished with the experiment, place the unreacted manganese and zinc in the appropriate strainers (in the sink). These metals can be reused.

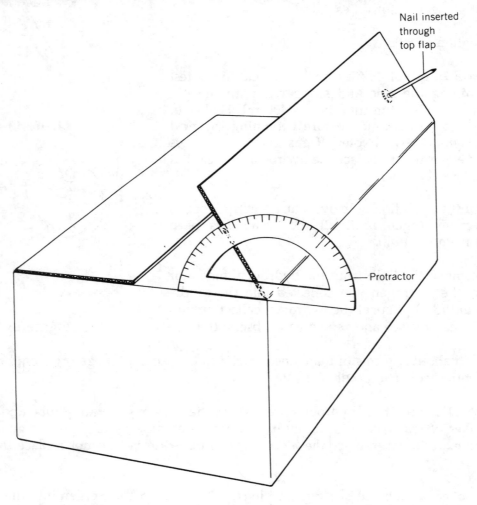

Figure 13-6

Prelaboratory Exercises - *Experiment 13*

For solutions to the starred problems, see *Appendix A*.

Name_____Date_____Lab Section_____

1.* The Haber process for the preparation of ammonia involves a combination reaction between nitrogen and hydrogen.

   a. Write a balanced molecular equation for the reaction.

   b. Give the optimum nitrogen to hydrogen mole ratio for the reaction.   _____

2. Gasoline is a mixture of many compounds including isomers of octane (C_8H_{18}). The products of the complete combustion of gasoline are water and carbon dioxide.

   a. What are isomers? Give an example of two isomers.

   b. Write a balanced reaction for the combustion of octane.

   c. Give the optimum octane to oxygen mole ratio for the reaction.   _____

   d. Write a balanced equation for the reaction of octane to give water and carbon monoxide.

   e. Explain why the presence of insufficient oxygen in a car cylinder could lead to the production of carbon monoxide and why this is a very undesirable outcome.

3. Heating potassium chlorate results in decomposition to potassium chloride and oxygen gas. This reaction is sometimes used as a method for generating oxygen.

   a. Write a balanced equation for the decomposition of potassium chlorate.

   b. Calculate the mass in grams of oxygen generated by the decomposition of 2.00 grams of potassium chlorate.

   _____

   c. Assume that the oxygen in 3-b is produced at 22°C and 760 torr pressure and use the ideal gas law to calculate the volume of oxygen produced.

   _____

4. The densities of methane, air and oxygen are approximately 0.66, 1.20 and 1.33 g/L respectively at 20°C. Should bottles of methane be stored mouth up or mouth down? Explain your answer.

5. For the water displacement method to be viable for gas collection, the gas must have certain properties.

   a.* What are these properties.

   b. Which of the following gases can be collected by water displacement: N_2, HCl, NH_3? Explain your answer.

Results and Discussion - *Experiment 13*
CHEMICAL PROPERTIES OF OXYGEN AND HYDROGEN

Name_____Date_____Lab Section_____

A. The candle flame.

1. Observations (include comments on states of matter, physical and chemical changes):

2. What do you think is burning? Explain your answer.

3. Further observations:

   a. match + solid wax -

   b. match + liquid wax -

   c. match + waxless wick -

   d. beaker partially over burning candle -

   e. stirring rod next to extinguished wick (hint: this is probably the key observation) -

   f. lighted match above extinguished flame

4. Was your first explanation *(question A-2)* correct or do you want to modify it or suggest a new one? Explain your answer.

5. Paraffin wax has the approximate formula $C_{25}H_{52}$ and burns to form carbon dioxide and water. Write a balanced equation for the burning of wax. Are your observations in *3-d* above consistent with this equation?

B. Preparation and properties of oxygen.

1. Combustion of candle   <u>Time (seconds)</u>

   a. Bottle 1   _____

   b. Bottle 3   _____

   c. Air   _____

2. Do your results indicate that the contents of bottles 1 and 3
   were the same? If not, account for your observations.   _____

3. Account for the difference in combustion time between bottle 3 and
   the air bottle.

4. Is there any oxygen left in the bottle when the candle flame goes out?
   Explain your answer.   _____

5. Did the wooden splint behave differently in the bottle of oxygen
   than in the air bottle? How?   _____

6. Diffusion of gases:

    a. Combustion of candle: <u>Time (seconds)</u>

        Top bottle (originally oxygen)   _____

        Bottom bottle (originally air)   _____

        Air bottle (copy from # *B-1c* above)   _____

        Oxygen bottle (copy from # *B-1b* above)   _____

    b. Compare and explain your observations for the top and bottom bottles. What conclusions can you come to about the oxygen content of each bottle?

    c. Assuming that the candle burned equal times in the top and bottom bottles (within experimental error), it is possible to show that each bottle contained about 60% oxygen. Show how to perform the calculation.

7. What did you observe when the charcoal was thrust into the fifth oxygen bottle?

8. What did you observe when the steel wool was thrust into the sixth oxygen bottle?

C. **Preparation and properties of hydrogen.**

    1. Write a balanced chemical equation for the reaction between zinc and hydrochloric acid.

    2. What did you observe when the hydrogen was "squirted" into the flame?

3. Write a balanced chemical equation for the reaction of hydrogen in the flame.

4. Write a balanced chemical equation for the decomposition of hydrogen peroxide into oxygen and water (The surface of manganese had air oxidized prior to its use to manganese dioxide. MnO_2 catalyzes the decomposition of H_2O_2 but do not include mangangese or manganese dioxide in the balanced equation.).

5. What did you observe when the oxygen was "squirted" in to the flame? Explain your observation.

6. Rate the report from each of the hydrogen - oxygen mixtures on a loudness scale of 1 to 10 with 10 being extremely loud.

hydrogen/oxygen ratio	rating	hydrogen/oxygen ratio	rating
_____	_____	_____	_____
_____	_____	_____	_____
_____	_____	_____	_____

7. What mixture seemed to give the loudest reports? Is this result consistent with the ratio predicted by the balanced equation (#C-3)? Explain your answer.

8. Describe the results of your rocket launch. What parameters affect the flight distance? What are the optimum values of these parameters?

9. Suggest any ways you can think of to improve any part(s) of this experiment.

Experiment 14

GAS LAWS

Learning Objectives

Upon completion of this experiment, students should have learned:
1. The validity of Boyle's Law.
2. Some aspects of graphing.
3. Dalton's Law of partial pressures.
4. The Ideal Gas Law.
5. Applications of stoichiometry.

Text Topics

Gas laws, stoichiometry (Malone, Chapters 9)

Comments

The first part of this experiment is a demonstration to be conducted with the class by the instructor. For additional problem solving experience involving the gas laws, see *Exercises 18, 19*.

Discussion

Although he probably wasn't the first person to prepare oxygen, Robert Boyle was the first known person to publish an account of its preparation. However, Boyle didn't prepare oxygen to show that it differed from other gases. He prepared it to show that it acted the same as all other gases under the influence of pressure. Before you read on, consider how you would expect the volume of a gas to change as the pressure of the gas is changed. You probably concluded that as the pressure goes up, the volume should decrease and vice versa. Consistent with this expectation, in 1662, Boyle found that the pressure on a gas is inversely proportional to its volume, or in equation form, $PV = a$ constant. This also implies that if pressure is altered on a confined volume of gas, the initial pressure times the initial volume equals the final pressure times the final volume, or $P_1V_1 = P_2V_2$. This equation is known as "Boyle's Law" and holds true as long as the temperature is held constant and the pressure isn't extremely great.

In 1787, Jacques Charles investigated the behavior of the volume of a gas as a function of temperature. Again, before you read on, try to predict what would happen to the volume of a gas as its temperature is raised or lowered. You probably concluded that an increase in temperature should result in an increase in volume and vice versa. Consistent with these expectations, Charles found that the volume of a gas kept at constant pressure varies directly with the Kelvin temperature, or in equation form: V/T = a constant. This implies that if the temperature of a confined gas is changed at constant pressure, the initial volume divided by the initial temperature equals the final volume divided by the final temperature, or:

$$\frac{V_1}{T_1} = \frac{V_2}{T_2}$$ This equation is known as "Charles' Law."

Using algebraic manipulation, it is possible to combine Boyle's and Charles' Laws into the so-called "Combined Gas Law."

$$\frac{V_1 P_1}{T_1} = \frac{V_2 P_2}{T_2} \qquad \text{or} \qquad V_2 = V_1 \times \frac{P_1}{P_2} \times \frac{T_2}{T_1}$$

The combined gas law is generally more applicable than Boyle's or Charles' Laws and is identical to Boyle's Law when the temperature is constant or Charles' Law when the pressure is constant. Even more general than the combined gas law is the ideal gas law, $PV = nRT$. Note that consistent with the earlier equations, the volume is proportional to the absolute temperature and inversely proportional to the pressure. The ideal gas law includes the additional dependence of the volume on the amount of the gas, expressed in this equation in moles (n). The proportionality constant, R, is called the gas constant.

In the first part of this experiment, you will attempt to verify Boyle's Law. The second part of this week's experiment combines the concepts of gas laws and stoichiometry. It involves the study of the stoichiometry of a gas producing reaction. Active metals such as sodium, potassium, calcium, magnesium, tin, and zinc react with acids to produce hydrogen gas. For example:

$$Zn_{(s)} + 2\,HCl_{(aq)} = ZnCl_{2(aq)} + H_{2(g)}$$

This reaction is often called a single replacement reaction or a redox reaction. As one of the products is a gas (H_2), it is possible to calculate the number of moles of hydrogen produced from an experimental determination of the volume of hydrogen using the ideal gas law, $PV = nRT$, or $n = PV/RT$. Once the number of moles of hydrogen is known, it is possible to calculate other important quantities. In today's experiment you will use the available data to calculate the atomic mass of zinc. The stoichiometry of the reaction tells us that it takes one mole of zinc to produce one mole of H_2 (assuming as will be the case in today's experiment that an excess of HCl will be used). As you will weigh the amount of zinc to be used and the number of moles of zinc can be calculated from the volume of H_2 produced, it will be possible to calculate the atomic mass of zinc from M = mass/moles.

In *Part C* of this experiment, you will determine the density of natural gas. From the density (d), you will calculate the average molecular mass of natural gas. Substituting the mass (m) divided by the molecular mass (M) for n (moles) in the ideal gas law yields $PV = \dfrac{mRT}{M}$.

Thus $M = \dfrac{mRT}{VP} = \dfrac{dRT}{P}$.

Procedure

A. **Boyle's Law, a demonstration.** In order to demonstrate Boyle's Law, a Boyle's Law apparatus is needed. This can be built using a 50 milliliter pinchcock buret, a meter stick, a ringstand, a ring, a leveling bulb, several feet of flexible rubber tubing, a rubber stopper and about 250 milliliters of mercury. See *Figure 14-1*.

Record the barometric pressure. Adjust the leveling bulb so the mercury in it is the same height as in the buret.

1. Using the meter stick as a reference, record the height of the top of the air column and the height of the mercury in the buret.

2. Now raise the leveling bulb to the highest place where the mercury level in it can be read on the meter stick, and record this height. Also record the height of the mercury in the buret. Now take several sequential readings by lowering the bulb about 15 cm each time.

The buret is cylindrical and the volume of a cylinder equals $\pi r^2 L$. Since r^2 and π are constant, the volume of an air column in a buret is proportional to its length. Therefore, in the calculations, the volume of air can replaced by the length of the air column.

The primary purpose of this experiment is to test Boyle' Law (e.g., does $P \times V =$ a constant at constant T?). By measuring the volume, V, at several pressures, you will be able to mathematically test Boyle's Law. After calculating the product PV, you should be able to determine if the results are constant within experimental error. An alternative test of Boyle's Law is the graphical method. One way to test data graphically is to plot the data in a way that a straight line should result. Boyle's Law can be rewritten in the form $V = m/P$ where m is a constant. As the general equation for a straight line is $y = mx + b$, a plot of V versus P should not yield a straight line. If, instead, you plot V vs 1/P, ($y = V$, $x = 1/P$ and $b = 0$), a straight line should result.

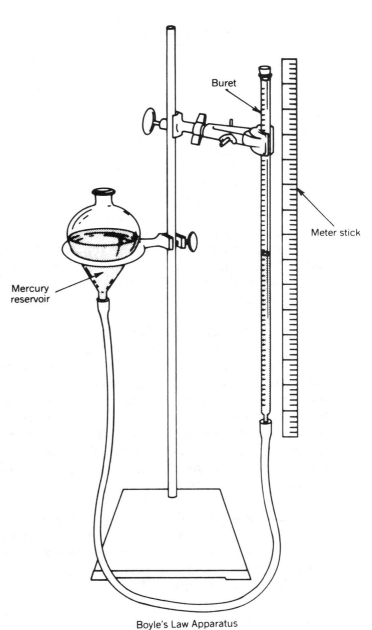

Boyle's Law Apparatus

Figure 14-1

B. Atomic mass of zinc. (*Caution: Flames should be kept strictly away from the flasks containing hydrogen and air.*) Set up the apparatus pictured in *Figure 14-2*. Fill the 500 mL Florence flask (or Erlenmeyer flask) up to the neck with water and put 25 mL of 6 M hydrochloric acid into the 250 mL Erlenmeyer flask.

Figure 14-2

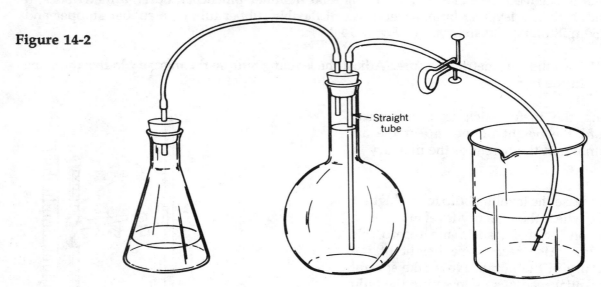

Straight tube

While the Erlenmeyer flask is *unstoppered,* apply compressed air (or use a rubber bulb) to the tube inserted through the stopper. This will cause water to flow from the Florence flask into the beaker. As soon as the flow begins, clamp the tube through which the water is flowing while disconnecting the compressed air source. Now that the delivery hose is filled with water, unstopper the 500 mL Florence flask, return the water you collected in the beaker to it and restopper it.

Weigh a little less than a gram of zinc metal to at least the nearest 0.01 gram. Holding the metal sample with one hand and the stopper for the 250 mL Erlenmeyer in the other, drop the sample into the HCl, stopper the flask, and remove the clamp. When the metal has completely reacted, allow a couple of minutes for the gas in the flasks to cool to room temperature. Equalize the pressure between the Florence flask and the beaker by raising the beaker or flask (whichever liquid level is lower) so that the level of the water you have collected in it is the same as the level of water in the flask. *Do not unstopper the flasks until after you have measured the water!* Clamp the delivery hose when the water levels are equal. Measure the amount of water you collected using a 500 mL graduated cylinder and also measure the temperature of the water.

Because you collected the hydrogen over water, part of the gas in the Florence flask was water vapor. Consequently, to obtain the actual pressure of hydrogen that was produced, you must subtract the vapor pressure of the water from the pressure at which the gas was collected (atmospheric). The answer sheets on page 170 will take you through the steps necessary to calculate the atomic mass of zinc.

Temperature (°C)	Vapor Press. (mm)	Temperature (°C)	Vapor Press. (mm)
10	9.2	21	18.6
11	9.8	22	19.8
12	10.5	23	21.1
13	11.2	24	22.4
14	12.0	25	23.8
15	12.8	26	25.2
16	13.6	27	26.7
17	14.5	28	28.3
18	15.5	29	30.0
19	16.5	30	31.8
20	17.5		

C. Density of natural gas. Stopper a clean dry 250 mL Erlenmeyer flask, weigh it and record the mass to the nearest milligram. In a good hood, unstopper the flask and fill it with natural gas. To do this, hold the flask upside down, insert the gas hose and run the gas moderately rapidly into it. After 30 seconds to a minute, drop the hose from the flask and immediately stopper the flask tightly. Turn off the gas valve, carry the flask upside down to the balance and weigh it to the nearest milligram.

Fill the flask completely to the top with water and stopper it (over the sink) as tightly as you had when collecting the gas. Unstopper it and measure the volume of water (the original volume of the gas) in a 500 mL graduated cylinder.

To determine the mass of gas in the flask, you must first determine the mass of the stopper and flask *without the air in it*. The mass of air originally in the flask equals its density multiplied by its volume. Find the density of the air in the laboratory from the density of the dry air chart in the CRC *Handbook of Chemistry and Physics*. You will need to know today's barometric pressure and the temperature. The mass of the natural gas in the flask can now be calculated by performing the appropriate subtractions.

The density of natural gas can now be calculated by dividing its mass by its volume. From its density, the molecular mass of natural gas ($M = dRT/P$) can be calculated and compared to the molecular mass of its most abundant constituent, methane (CH_4).

Chemical Capsule

In 1939, Otto Hahn and Fritz Strassman with the important assistance of Lise Meitner (it has been recommended that element 109 be named Meitnerium - Mt - in recognition of the contributions Lisa Meitner made to our knowledge of nuclear chemistry) demonstrated that Enrico Fermi had performed the first nuclear fission reaction in 1938. Fermi did not analyze the products of the reaction and did not realize that he had observed nuclear fission. It is now known that only a few isotopes of heavy elements (including $^{235}_{92}U$ and $^{239}_{94}Pu$) undergo nuclear fission when impacted by neutrons. One of the possible nuclear fission reactions for $^{235}_{92}U$ is:

$$^{235}_{92}U \ + \ ^{1}_{0}n \ \rightarrow \ ^{142}_{56}Ba \ + \ ^{91}_{36}Kr \ + \ 3\,^{1}_{0}n$$

It is important to notice two features of the above reaction. First, in addition to the products given, a huge amount of energy is evolved with this reaction. Second, the three neutrons produced can go on to initiate three more fission reactions. Thus, if there is sufficient $^{235}_{92}U$ (called the critical mass) to capture more than one neutron per fission reaction, a chain reaction can occur. This is the basis of an atomic bomb.

As you should be able to surmise from the atomic mass of uranium in the periodic table, not much of naturally occurring uranium is $^{235}_{92}U$. Uranium in nature consists of 99.27% $^{238}_{92}U$, 0.72% $^{235}_{92}U$ and 0.01 % $^{234}_{92}U$. When impacted by neutrons, $^{238}_{92}U$ does not fission but captures the neutron and then quickly undergoes beta decay.

$$^{238}_{92}U \ + \ ^{1}_{0}n \ \rightarrow \ [^{239}_{92}U] \ \rightarrow \ ^{239}_{94}Pu + \ ^{0}_{-1}e$$

In $^{235}_{92}U$ atomic bombs and nuclear energy facilities, the reaction above prevents the chain reaction from occurring. To sustain a nuclear fission reaction, natural uranium must be considerably enriched in $^{235}_{92}U$. During World War II, the leaders of the United States decided to attempt the construction of an atomic bomb (Manhattan Project). One of the problems that confronted the scientists involved in the Manhattan Project was how to provide uranium rich enough in $^{235}_{92}U$ to sustain a chain reaction.

This was cleverly accomplished by taking advantage of the very small difference in the diffusion rates of the two isotopes of uranium. Uranium was converted to uranium hexafluoride (UF_6) which has a diffusion rate ratio for the two isotopes of 1.0043. UF_6 is converted to a gas at very mild temperatures and then passed many times through diffuser tanks until a sufficient concentration of $^{235}_{92}U$ is attained.

Prelaboratory Exercises - *Experiment 14*

For solutions to the starred problems, see *Appendix A*. For additional problem solving experience involving the gas laws, see *Exercises 18, 19*.

Name_____Date_____Lab Section_____

1.* Convert a pressure of 0.90 atm to mm_{Hg}.

2.* The pressure on a 75 mL volume of gas is increased from 730 mm to 780 mm while constant temperature is maintained. What is the new volume of the gas?

3. A 1.5 L balloon containing air at 22°C is immersed in liquid nitrogen (boiling point = -196°C).

   a. Assume that the stretching forces of the rubber balloon can be neglected, calculate the predicted volume of the balloon at -196°C.

   _____

   b. When this experiment is performed, the volume is much smaller than the value calculated in *3-a*. Explain this observation.

4. A gas is collected over water at 20°C at a pressure of 752 mm. What is the pressure of the pure gas?

5.* When a 0.500 gram sample of magnesium reacts with excess HCl, a volume of 548 mL of hydrogen at a corrected pressure of 735 mm at 25°C is obtained. What is the experimental value for the atomic mass of magnesium and the percent error?

6.* 1.32 gram of an unknown gas occupies 0.750 L at 750 mm pressure and 27°C. What is the molecular mass of the gas?

7. Typical popcorn kernels weigh about 0.13 g, have a volume of 9.5×10^{-2} mL and contain about 1.7×10^{-2} g of water.

   a. Use the ideal gas law to calculate the pressure of gaseous water at 100°C confined to the kernal produced from the conversion of the water to gas .

   _____

   b. Explain why a popcorn kernel pops.

8. Ammonium nitrate is very valuable as a fertilizer because of its high nitrogen content. Unfortunately this also makes it potentially explosive. In 1947, ammonium nitrate on a ship in Texas City exploded killing 600 people and injuring 3500.

   a. Write a balanced reaction for the decomposition of ammonium nitrate into nitrogen, water and oxygen.

   b. For each mole of ammonium nitrate that decomposes, how many moles of gas are produced (because of the exothermicity of the reaction, the water is produced is in the gas phase).

   c. The density of ammonium nitrate is 1.72 g/cm³. Assuming the products are produced at 100°C and 1.00 atm pressure, calculate the ratio of the volume of gas produced to the volume of ammonium nitrate used.

Results and Discussion - *Experiment 14*
GAS LAWS

Name_____Date_____Lab Section_____

A. Boyle's Law

1. Atmospheric pressure (P_o)   _____

2. Top of air column (A)   _____

trial #	buret level B (mm)	Volume* V = A - B (mm)	Reservoir level C (mm)	Pressure difference D (mm) = C - B	Pressure P = P_o + D (mm)	P × V (mm)^2	1/P (1/mm)
1							
2							
3							
4							
5							
6							
7							
8							
9							
10							

*This is not actually a volume but is the length of the air column which is directly proportional to the volume of confined air.

3. On the first piece of graph paper provided, plot V vs P.

4. On the second piece of graph paper provided, plot V vs 1/P.

5. Does your data support Boyle's Law? Explain your answer.   _____

6. Average the values you obtained for P × V and use this value to calculate the volume at 0.85 atm. _____

7. According to your graph of V vs 1/P, what is the volume at 0.85 atm.? _____

B. Atomic mass of zinc.

1. Mass of zinc _____

2. Volume of water collected _____mL _____L

3. Temperature of water collected _____°C _____K

4. Atmospheric pressure _____mm

5. Vapor pressure of water (from table) _____mm

6. Corrected pressure of hydrogen (#4 - #5) _____mm _____atm.

7. Moles of H_2, n = PV/RT (be sure V is in liters, T is in Kelvin, P is in atm., and R = 0.08206 L atm/mol K) _____

8. Moles of zinc (equals moles of H_2) _____

9. Experimental atomic mass of zinc (#1/#7) _____

10. Atomic mass of zinc from periodic chart _____

11. Percent error _____

12. If efficient hoods are available, it is possible to boil off the excess HCl and water from the reaction flask and leave $ZnCl_2$ in the flask. What mass of $ZnCl_2$ should be left in the flask in your experiment? _____

C. Density of a gas.

1. Barometric pressure   _____mm   _____atm.

2. Temperature   _____°C   _____K

3. Density of dry air (*Handbook of Chemistry and Physics*)   _____

4. Volume of flask (from graduated cylinder measurement)   _____

5. Mass of air (density of air × volume of flask)   _____

6. Mass of flask + air + stopper   _____

7. Mass of empty flask + stopper (#6 - #5)   _____

8. Mass of flask + gas + stopper   _____

9. Mass of gas (#8 - #7)   _____

10. Density of gas (#9/#4)   _____

11. Molecular mass of gas $=$ dRT/P (where T = temperature in Kelvin, P = pressure in atmospheres and d = density of gas and R = 0.08206 L atm/mol K.)   _____

12. Molecular mass of methane (CH_4)   _____

13. Percent difference between your measured molecular mass of natural gas and the molecular mass of methane   _____

14. Suggest any ways you can think of to improve any part(s) of this experiment.

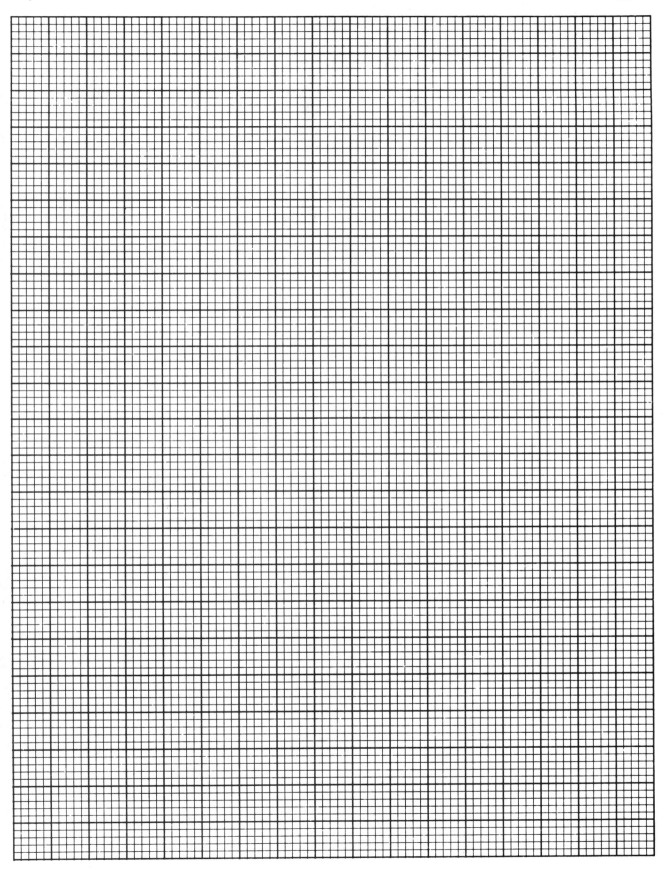

DISTILLATION AND HARDNESS OF WATER

Learning Objectives

Upon completion of this experiment, students should have learned:
1. To use distillation to purify a liquid.
2. To titrate.
3. Flame, conductivity and precipitation tests to analyze for water purity.

Text Topics

Vapor pressure, distillation, conductivity, solution stoichiometry, titrations, (Malone, Chapters 10, 11)

Comments

The distillation should be set up first and started. While the water is distilling, several of the tests can be performed on other water samples including the titration of tap water. This experiment will be even more interesting if you bring a water sample from home for testing. For additional problem solving experience with boiling points, see *Exercise 20*.

Discussion

In an earlier experiment *(Experiment 4)* , we utilized the technique of recrystallization to purify the solid, vanillin. Today's experiment deals with the most common purification method for liquids, distillation. Tap water will be distilled and then several of the properties of the original tap water will be compared to those of the distilled water and other water samples to determine the effectiveness of the distillation procedure. The principal analysis procedure will utilize a technique called titration to determine the hardness of the water.

You have undoubtedly heard water described as hard or soft. These seem to be strange terms to apply to a liquid. Can water really be "hard" in the classical sense of the word? It turns out that hardness has a different and special meaning when applied to water. Hardness is a measure of the combined amounts of calcium and magnesium salts present. As soap consists of fatty acid salts (e.g., sodium stearate) and the fatty acid salts of calcium and magnesium are insoluble in water, the effectiveness of soap decreases as hardness increases due to precipitation with magnesium and calcium. Titration is a very commonly used quantitative analysis technique and is easily applicable here to determine the total amount of calcium and magnesium ions in the aqueous solution. Other simple qualitative tests will be performed to check for the presence of sodium and chloride ions and conductivity will be determined as a measure of the total ion content of the water samples.

Distillation. Distillation is probably the most commonly used technique for the purification of liquids. Simply put, distillation involves energy input to convert a liquid to its vapor and then condensation of the vapor to liquid in a different part of the apparatus. Low boiling impurities will vaporize and condense first and can be collected and set aside while high boiling impurities will remain in the distilling flask unless more heat is provided. Boiling occurs when the vapor pressure of a substance equals the confining pressure. When the confining pressure is atmospheric pressure we call the temperature at which boiling occurs the normal boiling point. There are many parameters that can be changed to affect the efficiency of separation including the length and packing of the distilling column, the distillation rate, and the confining pressure. Today's distillation will be a simple distillation at atmospheric pressure.

Water softeners and ion exchange. Other methods of decreasing hardness involve replacement of calcium and magnesium ions by other ions. Home water softeners usually replace calcium and magnesium ions with sodium ions. Other cation exchange resins replace cations with hydrogen ions. Anion exchange resins usually replace anions with chloride or hydroxide ions. If hydrogen and hydroxide resins are used, hydrogen and hydroxide ions are produced which react with each other to yield water. Water purified by this method is called deionized water. This is an excellent technique for the purification of water and is now more commonly used to purify water for chemistry labs than distillation.

Complexiometric titration. The total calcium and magnesium concentration in water can be determined by titration with a solution of ethylenediamine-tetraacetic acid (EDTA) of known concentration. As the concentrations of calcium and magnesium are very low on a molarity scale, it is common practice to convert molarity to parts per million (ppm) when reporting hardness.

Other analyses. The total ion content of an aqueous solution is related to the conductance of the solution. Relative conductances of the water samples will be determined to compare the success of the different purification methods used. A flame test (see *Experiment 10*) will be used to qualitatively test for the presence of sodium ions, and a precipitation test will be used to test for insoluble silver salts.

Procedure

A. Distillation. [Note: Reread the **Comments** on page 175] Set up an apparatus similar to the one in *Figure 15-1*. Use flasks of approximately 150 mL in volume for the distilling pot and receiver. Fill the distilling flask about half full with tap water (never more than ⅔) and add a couple of boiling chips to the flask to prevent bumping. Be sure that the water flows uphill in the condenser (why?) and the mercury bulb of the thermometer is slightly lower than the junction of the condenser with the distillation column. Gently heat the distillation flask with a flame. Eventually boiling, condensation, and collection of the liquid will occur. During the distillation, it should be possible to set up and even partially perform some of the hardness titrations and the other water purity tests.

Discard the first milliliter (~20 drops) as it might contain volatile impurities and the impurities from the distillation glassware. Record the temperature in the distilling flask as soon as you have collected a milliliter of the distillate, then collect distillate until there is about 5 mL of liquid left in the distillation flask. Record the temperature in the distilling flask at the point you stop the distillation; then turn off the burner [Note: Never distill to dryness - it can sometimes lead to explosions]. Stopper the distillate and save it for testing.

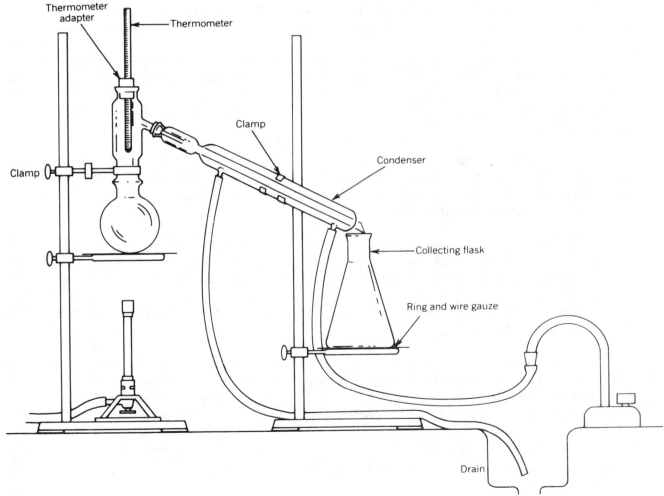

Figure 15-1

B. Water analysis. Water samples include tap water, 0.1 M NaCl, distilled water, laboratory deionized (or distilled) water and possibly home tap water.

1. <u>Conductivity</u>. While a simple LED conductivity device will suffice for this test, a commercial conductivity apparatus with a probe would be better [or the apparatus suggested by E. Vitz, *J. Chem. Ed.*, <u>64</u>, 628 (1987) with the quantification modification]. If time and the number of these instruments is limited, it is often expedient for the instructor to measure the conductances of the tap water, 0.1 M NaCl and laboratory deionized water and let students measure conductances of their distilled water sample. Some excitement can be added if a contest is held to see which student obtains the lowest conductance for his/her distilled water sample.

2. <u>Insoluble silver compounds.</u> Test for the formation of precipitates in all the water samples by adding a few drops of 0.1 M $AgNO_3$ to about 0.5 mL of each sample in 13×100 mm test tubes. [Note: Do this part only with the approval of the instructor as silver nitrate is expensive, toxic and a disposal problem.]

3. <u>Flame tests.</u> First clean a platinum or nichrome wire by alternately dipping it into 6 M HCl and holding it in a flame until no significant amount of color is observed. Put a few drops of each sample into clean test tubes and then test each water sample in the flame being sure to clean the wire between each sample. Do not put the wire into the original water samples as the HCl on the wire will contaminate the samples.

4. <u>Water hardness by titration.</u> Water hardness will be determined by buret titration on tap water and distilled water (and home tap water if available). Rinse and fill a clean buret with standardized 5.00×10^{-3} M disodium EDTA solution. Open the stopcock and allow the EDTA to displace the air in the buret tip. Close the stopcock and wipe off any liquid adhering to the tip of the buret.

Rinse a 25 mL pipet with tap water and pipet 25 mL of tap water into a 250 mL Erlenmeyer flask. Drain the pipet into the flask and touch the last drop off onto the interior wall of the flask. Do not blow the liquid that remains in the pipet tip into the flask. Doing so would give you more than 25.00 mL of liquid.

Add 20 drops of ammonia-ammonium chloride buffer and 2 drops of eriochrome Black T indicator to the flask. Read the buret to the nearest 0.01 mL and begin adding EDTA to the water sample. Slow down to dropwise addition at the first indication that the red color is changing to blue. When all the calcium and magnesium have reacted with EDTA, the red color will be replaced by a blue color. When the color changes to blue, stop the titration, read the buret and record the reading.

Rinse out your pipet with a little of your distilled water and pipet 25.00 mL of it into a clean but not necessarily dry flask (rinsing with deionized water is adequate). Add buffer and indicator as above and titrate with EDTA solution. [Hint: The distilled water should only require a few drops of EDTA solution before the endpoint is attained.]

Options: Rinse the pipet with your home tap water or an unknown supplied by the instructor and pipet the unknown into a clean flask. Add buffer and indicator and titrate with EDTA solution.

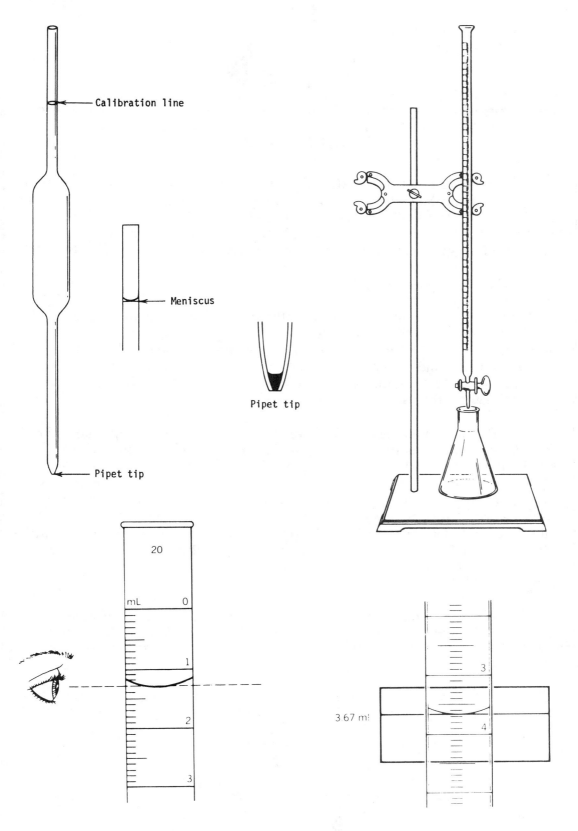

Figure 15-2

Chemical Capsule

Ethanol or ethyl alcohol (C_2H_6O) is recognized by most as the intoxicating compound in alcoholic drinks. In hard liquor such as vodka, ethanol is usually about 40% by volume (80 proof), wine consists of about 10% alcohol and beer is about 6% alcohol. Ethanol is produced by the anaerobic fermentation (in the presence of oxygen, acetic acid is produced instead of ethanol) of glucose by yeast:

$$C_6H_{12}O_6\text{(aq)} \quad \rightarrow \quad 2\ C_2H_6O\text{(aq)} \quad + \quad 2\ CO_2\text{(g)}$$

Fermentation yields a maximum of 15% ethanol because the ethanol denatures the yeast enzyme. The solution is distilled to achieve higher concentrations of ethanol. Surprisingly, even distillation yields a maximum of 95% ethanol. This is due to the fact that ethanol and water form a constant boiling mixture called an azeotrope that consists of 95% ethanol and 5% water. To obtain 100% ethanol from this mixture requires the use of other techniques.

Alcohols containing three or fewer carbons are miscible with water and can even be thought of as derivatives of water. If one of the hydrogens of water (HOH) is replaced with an alkyl group (C_nH_{2n+1}) such as ethyl, an alcohol results. The simplest alcohol, methanol (CH_4O) is called wood alcohol. In the body, methanol is oxidized to the very toxic chemicals, formaldehyde and formic acid. As little as 30 mL of methanol can be lethal. Smaller amounts can lead to blindness. Deaths have resulted from the substitution of methanol for ethanol. Ethanol is metabolized by an enzyme in our bodies called alcohol dehydrogenase to acetaldehyde. It is believed that acetaldehyde is at least partially responsible for hangovers.

Ethanol has many other uses in addition to alcoholic beverages. For example, it is used in organic synthesis, as a solvent and as a gasoline additive.

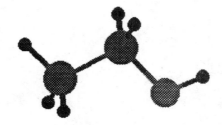

Prelaboratory Exercises - *Experiment 15*

For solutions to the starred problems, see *Appendix A*.

Name_____Date_____Lab Section_____

1. Define the terms below:

   boiling point

   vapor pressure

2. As the temperature of a liquid increases, does the vapor pressure
   increase or decrease?   _____

3. As the confining pressure on a liquid decreases, does the boiling point
   increase or decrease?   _____

4. The vapor pressures of several compounds are given below at 20°C. Rank them in terms
   of increasing boiling point with 1 the lowest and 6 the highest.

compound	vapor pressure (torr)	boiling point rank
acetone	162	_____
ethanol	41	_____
ether	380	_____
ethylene glycol	8.6×10^{-2}	_____
mercury	1.2×10^{-3}	_____
water	17.5	_____

5.*
   a. Atmospheric pressure in Denver, Colorado is typically about 0.84 atm. Refer to the
      Handbook of Chemistry and Physics to find the boiling point of water at this pressure.

                                                                  _____

   b. Will food cooked in boiling water cook slower or faster in Denver? Explain your
      answer.

                                                                  _____

   c. What is the function of a pressure cooker?

6.* A water sample gives a yellow flame test and yields a precipitate with silver nitrate. What conclusions can you come to from these observations?

7.* For titration with 5.00×10^{-3} M EDTA, the hardness in ppm (assume all hardness is due to $CaCO_3$) can be calculated from:

$$\text{hardness (ppm)} = \frac{\text{(mL of EDTA)} \times 500}{\text{mL of water sample}}$$

25.00 mL of a tap water sample is titrated with 5.00×10^{-3} M EDTA solution. 6.25 mL are required to reach the endpoint. What is the hardness of the water in ppm?

8. Is hard water harmful to drink? Explain your answer.

Results and Discussion - *Experiment 15*
DISTILLATION AND HARDNESS OF WATER

Name_____Date_____Lab Section_____

A. Distillation

1. Barometric pressure   _____

2. Boiling point of water at measured pressure
   according to *Handbook of Chemistry and Physics*
   edition_____ page_____   _____

3. Experimental boiling range for collected sample   _____

B. Water Analysis

1, 2, 3.

water sample	conductance	AgNO$_3$ results	flame test results
tap water			
0.1 M NaCl			
distilled			
deionized			
home tap			

a. What conclusions can you draw from your conductance measurements? Comment especially on the relative purity of the water samples.

b. Write balanced molecular and net ionic equations for the reaction of silver nitrate with sodium chloride.

c. What conclusions can you draw from your observations on the addition of silver nitrate to each water sample?

d. Did your flame test observations suggest the presence of any cations in any of the water samples? Explain your answer.

4. Hardness determinations

concentration of EDTA   _____

	tap water	distilled water	home tap or unk.
final buret reading (mL)			
initial buret reading (mL)			
vol. EDTA (mL)			
vol. water sample (mL)			
$[Ca^{2+} + Mg^{2+}]$[1] (mol/L)			
hardness[2] (ppm)			

[1]Multiply the molarity of EDTA by the volume used in Liters and divide by the volume in Liters of the water sample.
[2]See problem #7 in *Prelaboratory Exercises*

a. Did the distillation have a significant effect on the water hardness? Explain your answer.

b. Water with hardness in the range 0-60 ppm is termed soft, 60-120 ppm medium hard, 120-180 ppm hard and above 180 ppm very hard. Classify the water samples that you titrated.

c. Should there be and is there a correlation between the conductance and hardness results? Explain your answer.

d. Suggest any ways you can think of to improve any part(s) of this experiment.

Experiment 16

IONIC REACTIONS AND CONDUCTIVITY

Learning Objectives

Upon completion of this experiment, students should have learned:
1. The concepts of electrolytes and conductivity.
2. The driving forces and characteristics of double replacement reactions.
3. The fundamentals of formula and net ionic equation balancing.

Text Topics

Electrolytes, conductivity, solubility, double replacement reactions, net ionic reactions (Malone, Chapter 7, 11, 13)

Comments

The conductivity portion of this experiment has been designed as a demonstration by the instructor. It is also possible for students to perform this experiment with battery powered LED circuits. For additional problem solving experience with balancing formula and net ionic equations, see *Exercises 15, 16.*

Discussion

Sugar, salt, acetic acid, sodium hydroxide and hydrogen chloride are compounds that are very soluble in water. While the reasons for their high solubility are related to their high polarities, some of the properties of their aqueous solutions such as electrical conductivity are significantly different. This experiment will examine the electrical conductivity of aqueous solutions and the reactions of compounds that ionize in water.

When salts, strong acids (such as HCl) and strong bases dissolve in water, they dissociate almost completely into ions:

$$NaCl \quad \xrightarrow{H_2O} \quad Na^+ + Cl^-$$

$$K_2SO_4 \quad \xrightarrow{H_2O} \quad 2K^+ + SO_4^{2-}$$

$$HCl \quad \xrightarrow{H_2O} \quad H^+ + Cl^+$$

$$NaOH \quad \xrightarrow{H_2O} \quad Na^+ + OH^-$$

The positive ions are strongly attracted to the partial negative charge of the water's oxygen and the negative ions are attracted to the partial positive charge of the water's hydrogen. Thus we say that the ions are hydrated. Solutions which contain ions are capable of conducting electricity. Ions conduct electricity in solution because they carry charges. Positive ions migrate toward the negative cathode. When they come into contact with it, they gain electrons and are reduced. Negative ions migrate toward the positive anode. They give up their electrons and are oxidized when they come in contact with it. The net result is that electrons are transported through the solution by the ions and the circuit is completed.

When a compound dissolves in water and enables the solution to conduct electricity, the compound is called an electrolyte. A compound must satisfy two criteria to qualify as a strong electrolyte. First it must have substantial solubility and second it must break up into ions when it dissolves. Soluble salts, strong acids and bases fall into this category. It is difficult to predict solubility for ionic compounds and often it is necessary to refer to tables for this information. Almost all compounds that contain ions of the Group IA elements or the polyatomic ions ammonium, nitrate and acetate are very soluble and strong electrolytes. To determine solubility and the ability to conduct for other compounds, refer to the solubility table on page 399.

Some soluble compounds remain predominantly in the molecular form in water but do ionize to a small extent. Compounds in this category such as acetic acid and ammonia are poor conductors of electricity and are called weak electrolytes. Compounds such as sugar that dissolve in water but do not break up at all into ions do not conduct electricity and are called nonelectrolytes.

As discussed earlier, soluble salts and strong acids and bases dissociate into ions in aqueous solution. If two solutions of different ionic compounds are mixed, a solution with four ions will result with new combinations possible.

$$MX + NY \quad \xrightarrow{H_2O} \quad M^+ + X^- + N^+ + Y^-$$

Although MX and NY are soluble, the new combinations possible, $M^+ + Y^-$ or $N^+ + X^-$, *could* result in an energetically favorable change to the system yielding an *observable* reaction.

1. If either MY or NX is insoluble (crystal energetics more favorable than solvation energetics), a solid called a precipitate will form in the solution.

Formula equation: $\quad NaCl_{(aq)} + AgNO_{3(aq)} = AgCl_{(s)} + NaNO_{3(aq)}$

Net ionic equation: $\quad Ag^+ + Cl^- = AgCl_{(s)}$

Note that in the ionic equation, the ions Na^+ and NO_3^- are not included as they are not changed by the reaction. Overall, the *positive and negative ions have exchanged partners*. This type of reaction is often called a *double replacement* reaction or *metathesis* reaction.

2. Double replacement reactions also occur when the exchange of partners results in a gas (such as ammonia when the solution is heated) or a compound that decomposes into a gas (such as carbonic acid, H_2CO_3 or sulfurous acid, H_2SO_3).

a. $NH_4Cl(aq) + NaOH(aq) \xrightarrow{\Delta} NH_3(aq) + NaCl(aq) + H_2O(l)$

b. $Na_2CO_3(aq) + 2\ HNO_3(aq) = 2\ NaNO_3(aq) + H_2CO_3(aq)$

$H_2CO_3(aq) = H_2O(l) + CO_2(g)$

overall reaction:

$Na_2CO_3(aq) + 2\ HNO_3(aq) = 2\ NaNO_3(aq) + H_2O(l) + CO_2(g)$

3. A third cause of double replacement reactions is the formation of a very slightly ionized compound such as water or acetic acid. These reactions are usually exothermic (evolve heat) and can sometimes be detected by touching the reaction vessel.

$HNO_3(aq) + NaOH(aq) = H_2O(l) + NaNO_3(aq)$

$NaC_2H_3O_2(aq) + HNO_3(aq) = HC_2H_3O_2(aq) + NaNO_3(aq)$

Double replacement equation writing. The equation for the reaction between barium chloride and sodium phosphate is balanced as follows:

Step 1:   Write the correct formulas of the reactants. Join the positively charged ion of one reactant with the negatively charged ion of the other. Write a plus sign after the first product and then join the positively charged ion of the other reactant with the negatively charged ion of the first reactant.

Hint 1:   Write the positive ion first in each product.
Hint 2:   Do not carry over subscripts unless they are part of polyatomic ions. All polyatomic ion subscripts are carried over.

$BaCl_2 + Na_3PO_4 \rightarrow Ba_?(PO_4)_? + Na_?Cl_?$

Step 2:   Determine the charges of the reactant ions. The charge of each reactant compound is zero, so if you know the charge of one ion in the compound, you can determine what the charge of the other one is.

Example: $BaCl_2$

Barium is in group IIA and has a charge of +2 in its compounds; therefore, since the compound $BaCl_2$ has a charge of zero, each of the two chlorides must have a charge of -1. In double replacement reactions, the charge of each ion in the product is the same as it was as a reactant.

$$Ba^{2+}Cl_2^{1-} \; + \; Na_3^{1+}PO_4^{3-} \; \rightarrow \; Ba_?^{2+}(PO_4)_?^{3-} \; + \; Na_?^{1+}Cl_?^{1-}$$

The charges of the product compounds also must be zero. To find the proper subscripts:

1. Make the magnitude of the charge of the negative part of the formula, the subscript of the positive part.
2. Make the magnitude of the charge of the positive part of the formula, the subscript of the negative part.

$$Ba^{2+}PO_4^{3-} \; \rightarrow \; Ba_3(PO_4)_2$$

3. If both subscripts of either compound are divisible by any number, divide through by that number: $Ti_2^{4+}O_4^{2-}$ becomes TiO_2.

[Caution: If the subscript of any polyatomic ion is greater than 1, put the polyatomic ion formula in parentheses and the subscript outside the right parenthesis.]

The still unbalanced equation now has the following form:

$$?\; BaCl_2(aq) \; + \; ?\; Na_3PO_4(aq) \; \rightarrow \; ?\; Ba_3(PO_4)_2(s) \; + \; ?\; NaCl(aq)$$

Step 3: When the formulas of all compounds are correct (charges are zero), you are finished with subscripts. Do not write any more of them or adjust the ones you have! Now balance the equation using coefficients (numbers in front of formulas). A coefficient denotes either the number of moles or the number of whole formula units of the formula it precedes.

Find the ion with largest subscript (do not consider subscripts within a polyatomic ion). If there is a tie (as below with sodium and barium ions both having subscripts of 3), choose the ion with the largest oxidation state (barium ion). Imagine a 1 in front of the compound, barium phosphate and base the coefficients of all other compounds in the equation on barium phosphate. Work back and forth across the equation.

$$?\; BaCl_2(aq) \; + \; ?\; Na_3PO_4(aq) \; \rightarrow \; 1\; Ba_3(PO_4)_2(s) \; + \; ?\; NaCl(aq)$$

Now balance the barium ions on the left side of the equation by inserting a coefficient of 3.

$$3\; BaCl_2(aq) \; + \; ?\; Na_3PO_4(aq) \; \rightarrow \; 1\; Ba_3(PO_4)_2(s) \; + \; ?\; NaCl(aq)$$

The coefficient of 3 in front of $BaCl_2$ locked in 6 chlorides so a 6 is now needed in front of NaCl on the right.

$$3 \ BaCl_2(aq) \ + \ ? \ Na_3PO_4(aq) \ \rightarrow \ 1 \ Ba_3(PO_4)_2(s) \ + \ 6 \ NaCl(aq)$$

The 6 locks in 6 sodiums so the coefficient in front of Na_3PO_4 is the number of sodiums needed (6) divided by the number per formula unit (3) resulting in a coefficient of 2. Finally check to see if the phosphates are balanced to be sure you haven't made an error.

$$3 \ BaCl_2(aq) \ + \ 2 \ Na_3PO_4(aq) \ = \ Ba_3(PO_4)_2(s) \ + \ 6 \ NaCl(aq)$$

Ionic equations. To convert an equation from formula form to the total ionic form, separate all soluble ionic species with + signs and replace their subscripts with coefficients. Multiply the original coefficient by the subscript. (Soluble salts, strong acids and strong bases are totally ionic.) Retain subscripts for polyatomic ions, insoluble compounds, gases and slightly ionized compounds. Do not break up polyatomic ions into their component elements. Write molecular compounds that do not ionize in formula form (water should be written H_2O and not H^+ + OH^-. *Note that the term total ionic equation is misleading.* Be sure that not only the elements are balanced but the charges are also balanced.

When you convert the formula form of an equation into ionic form, leave gases, insoluble compounds and weakly ionized compounds in formula form.

$$3 \ BaCl_2(aq) \ + \ 2 \ Na_3PO_4(aq) \ = \ Ba_3(PO_4)_2(s) \ + \ 6 \ NaCl(aq)$$

This equation in total ionic form is:

$$3 \ Ba^{2+} \ + \ 6 \ Cl^- \ + \ 6 \ Na^+ \ + \ 2 \ PO_4^{3-} \ = \ Ba_3(PO_4)_2(s) \ + \ 6 \ Na^+ \ + \ 6 \ Cl^-$$

To convert a total ionic equation to a net ionic equation, cross out all species that don't change (these ions are often called spectator ions). What remains is the net ionic equation.

$$3 \ Ba^{2+} \ + \ 2 \ PO_4^{3-} \ = \ Ba_3(PO_4)_2(s)$$

Procedure

A. Conductivity. [Note: While we recommend that this section of the experiment be performed by the instructor as a demonstration experiment using a light bulb conductance apparatus or a large meter, it is also possible to perform it as a hands-on experiment using a simple apparatus that couples a 9 volt battery with an LED]. You or your instructor will determine the conductances of the solutions given in the *Results and Discussion* section. Additional instructions are included in that section.

B. Double replacement reactions. The goal of this experiment is to give you experience observing double replacement reactions and writing equations. To make the experiment more interesting, it has also been designed as a challenge to your thinking and reasoning skills. Six bottles labeled only with letters A - F with the solutions listed below will be provided.

compound	concentration
K_3PO_4	1.0 M
H_2SO_4	1.5 M
Na_2CO_3	1.0 M
NaOH	3.0 M
$Mg(NO_3)_2$	0.1 M
$SrCl_2$	0.1 M

Mixtures of two of the solutions will either undergo a double replacement reaction or not react. By comparing the predicted reaction matrix that you will prepare before the laboratory to the observations you make when you experimentally test the 15 possible mixtures, you should be able to assign identities to each of the bottles.

Before you come to laboratory, write possible double replacement equations for the 15 mixtures. Decide if a reaction should occur by determining if either or both of the products is insoluble in water (see table on page 399), a gas or a compound that decomposes into a gas (for this experiment look for H_2CO_3 = CO_2 + H_2O or formation of ammonia), or an un-ionized compound (such as water). When no reaction is predicted write *NAR* for "no apparent reaction". Also write total and net ionic equations for the 15 mixtures. Record your predicted observation (e.g., white precipitate) in the prediction matrix.

If two solutions (A and B) evolve heat when mixed, one is probably an acid and the other a base. If one (say A) of the two reacts with another solution (C) and evolves a gas, C is probably a carbonate or bicarbonate, A the acid and B the base. By using this kind of logic, you should be able to deduce the identities of the solutions.

[Notes: The mixture of potassium phosphate and sulfuric acid is somewhat complex. Very careful observers will notice that this mixture evolves a small amount of heat as compared to another mixture that undergoes an easily detected temperature change. For the purposes of this level course, the phosphate - acid reaction can be treated as NAR or perhaps extra credit can be awarded for students that analyze it correctly. It should also be noted that strontium hydroxide has a marginal solubility but careful observers will notice formation of a slight amount of precipitate.]

To perform the experiment, put 5 test tubes in a rack and add about 2 mL (about 40 drops) of solution A to each test tube. Now add 2 mL of B to the first test tube, 2 mL of C to the second, and so on. Mix the contents of each test tube thoroughly and check for the formation of a precipitate evolution of a gas, odor or heat (by touching the outside of the test tube). Record your observations in the experimental observation matrix. Now put four clean test tubes in the rack and add 2 mL of B to each. Test the solutions with the remaining solutions (C - F). Repeat this procedure until the 15 possible mixtures have been tested.

Compare the prediction matrix to the observation matrix and determine the identities of the solutions.

Prelaboratory Exercises - *Experiment 16*

For solutions to the starred problems, see *Appendix A*. For additional problem solving experience with balancing formula and net ionic equations, see *Exercises 15, 16*.

Name_____Date_____Lab Section_____

1.* Which of the following should be electrolytes? Mark *E* (electrolyte) or *N* (nonelectrolyte).

potassium chloride(aq)   _____

nitric acid(aq)   _____

potassium hydroxide(aq)   _____

silver chloride(aq)   _____

sodium sulfate(aq)   _____

calcium carbonate(aq)   _____

isopropyl alcohol(aq) ($CH_3CHOHCH_3$)   _____

acetone(aq) (CH_3COCH_3)   _____

2.* You have three bottles that contain 0.1 M $Ba(OH)_2$, 1 M Na_2CO_3 and 2 M HCl labeled in random order A, B and C. Write possible formula (FE), total ionic (TIE) and net ionic (NIE) equations for the three possible mixtures and fill in the expected observations in the prediction matrix. Given the observations included in the observation matrix, determine the identities of A, B and C.

a. barium hydroxide + sodium carbonate

FE _____

TIE _____

NIE _____

b. barium hydroxide + hydrochloric acid

FE _____

TIE _____

NIE _____

c. sodium carbonate + hydrochloric acid

FE _____

TIE _____

NIE _____

Na$_2$CO$_3$	HCl		
1	2	Ba(OH)$_2$	
	3	Na$_2$CO$_3$	

	B	C	
	heat	gas	A
		white ppt.	B

A _____

B _____

C _____

3. Fill in the predictions in *A-1* to *A-14* and *A-21, A-22* and *A-23*.

4. Write the formula equations, total ionic equations and net ionic equations and fill in the prediction matrix for *Part B*.

Results and Discussion - *Experiment 16*
IONIC REACTIONS AND CONDUCTIVITY

Name_____Date_____Lab Section_____

A. Conductivity of solutions (Mark SE, FE, WE, or NE for strong, fair, weak or non electrolyte)

#	Solution	Prediction	Observed Conductivity	
1.	0.1 M NaCl	_____	_____	
2.	1×10^{-2} M NaCl	_____	_____	
3.	1×10^{-3} M NaCl	_____	_____	
4.	crystalline NaCl	_____	_____	(go to #21)
5.	deionized water	_____	_____	
6.	tap water	_____	_____	
7.	0.1 M sucrose	_____	_____	
8.	0.1 M ethanol	_____	_____	
9.	0.1 M NaOH	_____	_____	
10.	0.1 M NH_3	_____	_____	
11.	0.1 M HCl	_____	_____	
12.	0.1 M $HC_2H_3O_2$	_____	_____	(go to # 22)
13.	0.5 M H_3PO_4	_____	_____	
14.	saturated $Ca(OH)_2$	_____	_____	(go to # 23)

To answer exercises 15 - 20 below, compare, contrast and explain your observations for the indicated exercises.

15. #1, 4

16. #1, 2, 3

17. #1, 5, 6

18. #1, 7, 8

19. #1, 9, 10

20. #1, 11, 12

For systems 21 - 23, predict what your observations will be when the instructions are followed. Perform the experiment, record and explain your observations writing net ionic equations when appropriate.

21. While monitoring the conductivity of initially crystalline NaCl, add water dropwise with stirring until all of the NaCl has dissolved.

Prediction:

Observation:

Explanation and net ionic equation:

22. While monitoring the conductivity of 0.1 M $HC_2H_3O_2$, add 0.1 M NH_3 dropwise with stirring until an approximately equal volume of 0.1 M NH_3 has been added.

Prediction:

Observation:

Explanation, formula and net ionic equations:

23. Add 3 drops of phenolphthalein indicator to 25 mL of saturated calcium hydroxide solution in a 150 mL beaker. While monitoring the conductivity of the calcium hydroxide solution, add 0.5 M H_3PO_4 dropwise with stirring (preferably with a magnetic stirrer).

Prediction:

Observation:

Explanation, formula and net ionic equations:

B. **Double replacement reactions.** Six unlabeled bottles containing the solutions below in some scrambled sequence will be provided.

   1 M K_3PO_4 3 M NaOH

   1.5 M H_2SO_4 0.1 M $Mg(NO_3)_2$

   1 M Na_2CO_3 0.1 M $SrCl_2$

For the 15 possible mixtures of the six solutions, write formula (FE), total ionic (TIE) and net ionic (NIE) equations. When no reaction is expected, write *NAR* for no apparent reaction. Based on these equations, fill in the prediction matrix. Based on your experimental observations, fill in the observation matrix. Compare the two and assign identities to *A - F*. (For solubilities, see *Appendix C* - page 399)

1. potassium phosphate + sulfuric acid

   FE_____

   TIE_____

   NIE_____

2. potassium phosphate + sodium carbonate

   FE_____

   TIE_____

   NIE_____

3. potassium phosphate + sodium hydroxide

   FE_____

   TIE_____

   NIE_____

4. potassium phosphate + magnesium nitrate

   FE_____

   TIE_____

   NIE_____

5. potassium phosphate + strontium chloride

FE_____

TIE_____

NIE_____

6. sulfuric acid + sodium carbonate

FE_____

TIE_____

NIE_____

7. sulfuric acid + sodium hydroxide

FE_____

TIE_____

NIE_____

8. sulfuric acid + magnesium nitrate

FE_____

TIE_____

NIE_____

9. sulfuric acid + strontium chloride

FE_____

TIE_____

NIE_____

10. sodium carbonate + sodium hydroxide

FE_____

TIE_____

NIE_____

11. sodium carbonate + magnesium nitrate

FE_____

TIE_____

NIE_____

12. sodium carbonate + strontium chloride

FE_____

TIE_____

NIE_____

13. sodium hydroxide + magnesium nitrate

FE_____

TIE_____

NIE_____

14. sodium hydroxide + strontium chloride

FE_____

TIE_____

NIE_____

15. magnesium nitrate + strontium chloride

FE_____

TIE_____

NIE_____

Prediction Matrix

	H₂SO₄	Na₂CO₃	NaOH	Mg(NO₃)₂	SrCl₂	
	1	2	3	4	5	K_3PO_4
		6	7	8	9	H_2SO_4
			10	11	12	Na_2CO_3
				13	14	NaOH
					15	$Mg(NO_3)_2$

Column headers: H_2SO_4, Na_2CO_3, NaOH, $Mg(NO_3)_2$, $SrCl_2$

Observation Matrix

	B	C	D	E	F	
						A
						B
						C
						D
						E

label color _____

A = _____   D = _____

B = _____   E = _____

C = _____   F = _____

Experiment 17

ANALYSIS OF CATIONS

Learning Objectives

Upon completion of this experiment, students should have learned:
1. To use a centrifuge and wash precipitates.
2. To determine the identities of unknown ions using a flow scheme.

Text Topics

Balancing equations, net ionic equations (Malone, Chapters 7, 11)

Comments

For additional problem solving experience with balancing equations and net ionic equations, see *Exercises 15, 16.*

Discussion

In the process of determining the formula of a hydrate in *Experiment 10,* you performed a simple series of tests to discover the identity of your inorganic salt. In *Experiment 16,* a chemical reactivity matrix enabled you to assign identities to six solutions. These two experiments were based on the assumption that only one chemical could be present. The paper chromatography experiment enabled you to analyze for four cations even when any of the four are mixed together. As discussed in the paper chromatography experiment, separation and analysis of mixtures is an important aspect of chemistry. Today's experiment will introduce you to a classical method of analyzing a mixture of cations. The procedure in some cases allows you to selectively test for ions in the presence of others and in other cases requires separation procedures before analysis. Two new techniques will be introduced, centrifuging and washing of precipitates.

For the purposes of this laboratory exercise, we will study a special group of ions selected primarily for educational and safety reasons. These ions are:

Na^+, NH_4^+, Sr^{2+}, Zn^{2+}

It is best to perform this experiment by running a "known" solution containing all of the possible ions alongside your unknown. In this way you can compare the reactions of the unknown to the characteristic reactions of each of the possible ions.

Procedure

Step 1: Divide your unknown into two equal portions. Label them "U_1" and "U_2". Put 3 milliliters of the known solution (containing all of the ions) into another test tube and label it "K_1". Add one drop of 6 M NH_3 to one of the portions of your unknown (U_1) and one drop to the known solution (K_1). Any precipitate that forms indicates the presence of zinc. Ammonia increases the concentration of hydroxide ions in solution and causes some insoluble hydroxides to precipitate. If a precipitate forms, centrifuge the mixture and save the decantate for the strontium test. If no precipitate forms, save the solution for the strontium test. Any precipitate may be discarded.

Step 2: Take the decantate or solutions to which you added the 6 M NH_3 (U_1 and K_1) and add 2 milliliters of 1.0 M $(NH_4)_2CO_3$ to each of them. Absence of a precipitate here indicates the absence of strontium, but a precipitate does not necessarily indicate its presence. A precipitate will form in K_1 but may or may not form in U_1. Centrifuge tubes in which a precipitate forms and discard the decantate. Wash the precipitate(s) by directing a stream of deionized water into the precipitate so that it is thoroughly agitated. Continue adding the water until the test tube is half full, then centrifuge and discard the wash water. Repeat the process. After two washings, add dropwise, just enough 6 M HCl to dissolve the precipitate(s).

Clean a platinum wire by dipping it into 6 M HCl and holding it in the burner flame until the flame returns to its original color. Dip the clean platinum wire into the solution to which you added the HCl and hold it in the Burner flame. A red flame indicates the presence of strontium. Now add 2 milliliters of 1 M Na_2SO_4 to the solution(s). A precipitate here confirms the presence of strontium. Absence of a precipitate indicates that strontium is absent.

Step 3: You are now finished with U_1 and K_1. You can pour them out and clean the test tubes. Put 2 milliliters of the known solution into a clean test tube. Label it K_2. Pour about half (or about one milliliter) of the known solution (K_2) into another clean test tube, clean your platinum wire and perform a flame test on this sample. The strong yellow flame that persists for one or two seconds is due to the presence of Na^+. Pour about a milliliter of your original unknown sample (U_2) into a clean test tube. Clean your platinum wire and perform a flame test on it. A strong yellow flame here confirms the presence of Na^+. Absence of the yellow flame indicates that your unknown does not contain Na^+. Since the test is very sensitive to traces of sodium, compare the intensity and duration of the flame you obtain with that from a sample of distilled water and that from 0.1 M NaCl solution. If strontium ion is present in the sample, you may also notice a red strontium flame.

Step 4: Add one milliliter of 6 M NaOH to the milliliter of known solution (K_2) you have left. Put the tube into a beaker of boiling water, moisten a piece of red litmus paper with deionized water and hold it directly over the mouth of the tube. Be careful that any liquid remaining on the lip of the test tube does not come into contact with the litmus paper. The litmus paper will gradually turn blue as it is exposed to the ammonia vapors being generated in the reaction between ammonium ion and sodium hydroxide. Now put a milliliter of your original untreated unknown (U_2) into a clean test tube. Add a milliliter of 6 M NaOH and put the tube into the boiling water bath. As you did with the known, hold a piece of red litmus that has been moistened with deionized water close to the mouth of the test tube. If ammonium ion is present in your sample, the litmus paper will gradually turn blue as it did with the known sample.

Prelaboratory Exercises - *Experiment 17*

For solutions to the starred problems, see *Appendix A*. For additional problem solving experience with balancing equations and net ionic equations, see *Exercises 15, 16*.

kkName_____Date_____Lab Section_____

      The most commonly encountered qualitative analysis scheme is used for an analysis of the "*Group I*" ions; Pb^{2+}, Ag^+ and Hg_2^{2+}. The *Group I* ions form the only commonly encountered insoluble chlorides and can be separated from most of the other cations by the addition of hydrochloric acid. The remainder of the *Group I* scheme then involves separation and analysis of the chlorides of lead(II), silver and mercury(I).

step

1   The three chlorides are precipitated by addition of HCl.

2   $PbCl_2$ dissolves in hot water so the solution is heated and filtered.

3   The filtrate is tested for lead(II) by adding K_2CrO_4 which precipitates the lead (II) as yellow $PbCrO_4$.

4   Ammonia is added to the washed precipitate. Ammonia forms a strong complex with silver ion, $Ag(NH_3)_2^+$, and the silver chloride dissolves. Mercury(I) chloride reacts to form the still insoluble $HgNH_2Cl$ + metallic mercury and the black residue confirms the presence of mercury.

5   The decantate from *step 4* is tested for silver ion by adding nitric acid. This neutralizes the ammonia and the silver recombines with the chloride to give back insoluble white silver chloride.

1.*   Write the net ionic equations (NIE) for each step.

step

1   ____Pb^{2+} + ____Cl^- =

     ____Hg_2^{2+} + ____Cl^- =

     ____Ag^+ + ____Cl^- =

2   ____$PbCl_2(s)$ =

3   ____Pb^{2+} + ____CrO_4^{2-} =

4   ____$Hg_2Cl_2(s)$ + ____$NH_3(aq)$ = ____$Hg(l)$ + ____$HgNH_2Cl(s)$ + ____NH_4^+ + ____Cl^-

5   ____$Ag(NH_3)_2^+$ + ____H^+ + ____Cl^- = ____$AgCl(s)$ + ____NH_4^+

2.* Draw and fill in a flow diagram for the *Group I* ions.

Flow diagram

Results and Discussion - *Experiment 17*
ANALYSIS OF CATIONS

Name_____Date_____Lab Section_____

1. Observations

	Known	Unknown
Step 1:		
Step 2:		
Step 3:		
Step 4:		

2. Finish the flow diagram below (for your known solution) by filling in the blanks for the reagents used for each reaction and by filling in the boxes with the appropriate cations or formula units. The codes (e.g., 2b) correspond to the net ionic equations in question # 2 above.

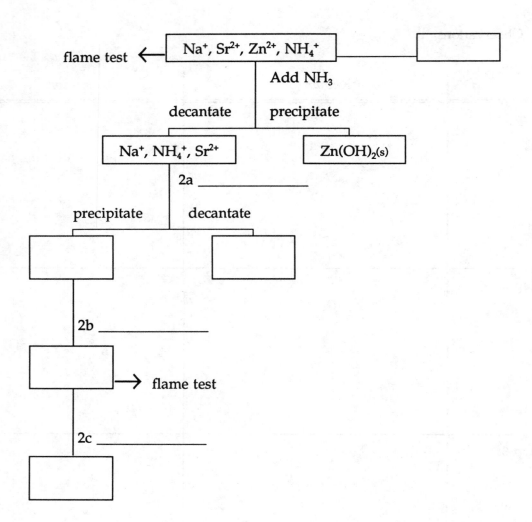

208

3. For the *known* solution, write net ionic equations for all reactions. Include steps for dissolving of solids as well as formation of precipitates. Remember that even in net ionic equations, solids are written as formula units.

Step 1:

1a ___Zn^{2+} + ___OH^- =

Step 2:

2a ___Sr^{2+} + ___CO_3^{2-} =

2b ___$SrCO_3(s)$ + ___H^+ =

2c ___Sr^{2+} + ___SO_4^{2-} =

Step 3:

4 ___NH_4^+ + ___OH^- =

4. Unknown number _____

    List the cations present in your unknown (0 - 4 possible) _____

    _____

    _____

    _____

    _____

Select the best answer for the following multiple choice questions:

5. Why is the centrifuge used rather than filtering to separate the liquid and solid phases in this experiment?   _____

    a. It is faster.
    b. It is more convenient for small amounts.
    c. The solid is too flocculent and will pass through the filter.
    d. The corrosive chemicals involved will dissolve the filter paper.
    e. a and b

6. Why is the precipitate washed with water before it is used for further testing?   _____

    a. To dissolve some of the precipitated solid.
    b. To clean the test tube.
    c. To remove the soluble salts.
    d. All of the above.
    e. None of the above.

7. In *Step 2*, $(NH_4)_2CO_3$ is added to precipitate the Sr^{2+} as $SrCO_3$. The appearance of the precipitate is not taken as conclusive evidence for the presence of strontium and further testing for strontium is performed. Why?   _____

   a.   One positive test is not enough.
   b.   Sodium could precipitate as sodium carbonate.
   c.   Any zinc ion remaining from *Step 1* could precipitate as $ZnCO_3$.
   d.   None of the above.

8. In *Step 4*, why is it important that the litmus not touch any of the liquid?   _____

   a.   Litmus reacts with water.
   b.   The sodium hydroxide poured down the side could turn the litmus blue if it touches it.
   c.   Litmus dissolves in liquid and only works with vapor.
   d.   None of the above.

9. Suggest any ways you can think of to improve any part(s) of this experiment.

Experiment 18

SPECTROSCOPY

Learning Objectives

Upon completion of this experiment, students should have learned:
1. To prepare a standardized solution using a volumetric flask.
2. Methods of quantitative dilution.
3. To use a Spectronic 20 or similar spectrometer.
4. Beer's law and its applications in analytical chemistry.

Text Topics

Concentration units, quantitative dilutions, spectroscopy (Malone, Chapters 5, 11)

Comments

For additional problem solving experience with solution concentrations, dilutions and the relationships between energy, frequency and wavelength, see *Exercises 9, 21.*

Discussion

This experiment has been designed to introduce you to two new techniques: quantitative dilutions of solutions and spectroscopy. Each group of students will prepare five $Co(NO_3)_2$ solutions from a 0.150 M $Co(NO_3)_2$ stock solution. The absorption of light at 510 nm for each solution will be measured. The data will be treated graphically and then the absorption of a $Co(NO_3)_2$ solution of unknown concentration will be measured and its concentration determined from the graph.

Spectroscopy is one of the most useful methods of analysis available in chemical and medical technology laboratories. Various types of spectroscopy are regularly used to identify compounds, determine molecular structure, identify what elements are present in samples, and determine the concentrations and amounts of specific elements or compounds in samples. During this course, you have probably already encountered some examples of the importance of spectroscopy. First, the observation that elements absorb or emit only specific and unique wavelengths of light (as you observed when performing flame tests) led Bohr, Schrodinger and others to formulate our present-day quantum theory.

You are probably also aware that some compounds and solutions can be distinguished on the basis of their colors (see *Experiment 6*). For example, a quick glance would enable you to differentiate between aqueous solutions of $SrCl_2$ and $CoCl_2$. The cobaltous ion absorbs green light (causing an electron to be promoted from one orbital to a higher one) and the solution therefore appears red to the eye. The $SrCl_2$ solution appears colorless which means that neither Sr^{2+} or Cl^- ions absorb energy in the visible region of the electromagnetic spectrum.

Examining light somewhat closer, we find light has characteristics of both a particle (photon) and a wave. As a result of its wave properties, light can be described in terms of its frequency (cycles per second) and its wavelength. As one might expect, the number of oscillations or cycles per second (frequency) is proportional to the energy of the light, $E = h\nu$ (E = energy, h = Planck's constant, ν = frequency). The higher the frequency, the higher the energy and vice versa. In addition, the speed of light (c) is the product of the frequency and the wavelength (λ), $c = \lambda\nu$. Therefore, the higher the frequency an electromagnetic wave has, the shorter its wavelength is. Substitution results in the conclusion that the energy is inversely proportional to the wavelength ($E = hc/\lambda$). Notice the very important fact that as the wavelength increases, the energy decreases. The table below gives some very approximate wavelengths and energies for commonly used regions of the electromagnetic spectrum. Notice the energy decreases in the sequence ultraviolet, visible and infrared and that blue is higher in energy than red. Also notice the relatively long wavelength of microwave energies. Contrary to some people's beliefs, microwaves do not "nuke" the food.

wave type	wavelength (nm)	energy (kJ/mol)	transition
gamma	10^{-1} - 10^{-3}	10^6 - 10^8	electronic (ionization)
X-rays	10 - 10^{-1}	10^4 - 10^6	electronic (ionization)
far ultraviolet	200 - 10	6×10^2 - 10^4	electronic (ioniz., promotion)
ultraviolet	400 - 200	3×10^2 - 6×10^2	electronic (promotion)
visible	700 - 400	1.7×10^2 - 3×10^2	electronic (promotion)
infrared	10^5 - 700	1 - 1.7×10^2	vibrational, rotational
microwaves, radar	10^8 - 10^5	10^{-3} - 1	rotational
radio, mag. res., TV	10^{12} - 10^8	10^{-7} - 10^{-3}	nuclear spin

Science has learned how to probe the structure of matter using a broad range of wavelengths of light. The table above gives you a general idea of the utility of the electromagnetic radiation spectrum. Note that the eye is sensitive to only a small portion of the spectrum. The nature of the light absorbed by a species can often help us identify it, as every species has a unique absorption spectrum. Additionally, the amount of absorption can enable us to determine the concentration of the species present. When light is absorbed by a sample, it is possible to derive a relationship between a defined quantity, the absorption (A); and the concentration and path length of the sample. This relation is called Beer's law. As you might have intuitively predicted, the absorption increases linearly as the concentration and pathlength are increased, $A = \varepsilon bc$ (ε = proportionality constant, b = path length, c = concentration of absorbing species). The proportionality constant, ε, is related to the probability that a species will absorb light at a specific wavelength.

Sr^{2+} for instance, has an ε of about zero throughout the visible region and is therefore a colorless ion. Co^{2+} on the other hand has an ε of about 5 at a wavelength of 510 nm and appears red to the eye. To determine concentrations using Beer's law, one first prepares a solution of the substance and determines its absorption spectrum. From the absorption spectrum, one selects the optimal wavelength for a study of the absorption versus wavelength. Usually this is the absorption maximum as it is the most sensitive wavelength. Next, as in today's experiment, one prepares a series of solutions of the substance at known concentrations. This is often done by making one solution of known concentration and quantitatively diluting it. Concentrations can then be calculated from $M_1V_1 = M_2V_2$. Absorption values (A) for the solutions are measured and plotted on a graph versus the concentrations. As ε and b are constant for a given substance with a fixed path length and wavelength, a straight line should result. If "A" values are then measured for solutions of the substance of unknown concentration, the graph can be used to read off the concentrations.

Procedure

A. Preparation of solutions. Prepare a 0.150 molar solution of cobalt nitrate by dissolving 2.18 grams of $Co(NO_3)_2 \cdot 6H_2O$ in about 25 mL of deionized water in a small clean beaker. Transfer the solution to a 50 mL volumetric flask. Use a directed stream of water from a wash bottle to transfer any cobalt nitrate solution clinging to the inside of the beaker into the volumetric flask. Then dilute to the mark with deionized water and *mix thoroughly* by inverting the tightly stoppered flask and shaking.

You will use a buret to prepare 5 mL each of five cobalt nitrate solutions of various concentrations from your 0.150 M cobalt nitrate stock solution. Number five clean, dry 13 × 100 mm test tubes or five Spectronic 20 cuvettes 1 through 5 near the top with a marker pen. Rinse either a 25 mL or 50 mL buret with three small portions (less than 5 mL each) of your cobalt nitrate solution. Transfer your stock solution to the buret (you will not have enough solution to fill a 50 mL buret but you will still have more than enough solution). Using the buret, deliver 1.00 mL of the cobalt nitrate solution into test tube number 1, 2.00 mL into #2, 3.00 mL into #3, 4.00 mL into #4 and 5 mL into #5. Rinse the buret several times with deionized water and fill it with deionized water. Now using the buret, deliver 4.00 mL of deionized water into test tube number 1, 3.00 mL into #2, 2.00 mL into #3 and 1.00 mL into #4. *Thoroughly mix the contents of each test tube.* Obtain a cobalt nitrate solution of unknown concentration from your instructor.

B. Beer's law graph. Using the appropriate dilution factors, calculate the concentrations of the $Co(NO_3)_2$ solutions and put the values in the table on page 219. Following the instructions below, measure the absorption of tubes 1 - 5 and the unknown at 510 nm. Use a tube with deionized water to zero the Spectronic 20.

214

Spectronic 20 operating instructions (refer to Figure 18-1):

1. The small red light to left of the meter indicates whether the instrument is on or not. To turn the instrument on, turn the left knob clockwise until you hear a click. Allow the instrument to warm up for 15 minutes.

2. Adjust the wavelength to 510 nm.

3. With the sample chamber empty, turn the left knob until the meter reads 0% transmittance (or A = ∞).

4. Fill a 13 × 100 mm test tube or cuvette with deionized water. Wipe it clean with a paper towel. Put it into the sample chamber and adjust the lower right knob until the meter reads 100% transmittance (A = 0.00).

5. Repeat numbers 3 and 4.

6. Remove the test tube containing deionized water from the sample chamber and replace it with a 13 × 100 mm test tube (wiped clean) containing one of the known concentrations of $Co(NO_3)_2$. Read the absorption value from the **bottom scale** on the meter.

7. Repeat number 6 for the four other known concentrations and the solution of unknown concentration.

8. Plot the absorption (y axis) versus the concentration (x axis) for the five numbered tubes and determine from the graph the concentration of cobalt nitrate in your unknown.

Note: If you are doing Option 2 of Part C, save the test tubes or cuvettes with solution #5 and with water.

C. Absorption spectrum of cobalt(II) nitrate.

Option 1 - the shorter, nonexperimental procedure. A 0.150 M $Co(NO_3)_2$ solution was prepared. Absorption values (A) for this solution were measured at several wavelengths. The data is given in the *"Results and Discussion"* section. Plot A (vertical axis) versus wavelength (horizontal axis).

Option 2 - Experimental determination. You will need the test tube or cuvette containing solution #5 and one with deionized water for this part.

Spectronic 20 operating instructions (refer to Figure 18-1):

1. The small red light to left of the meter indicates whether the instrument is on or not. To turn the instrument on, turn the left knob clockwise until you hear a click. Allow the instrument to warm up for 15 minutes.

2. Adjust the wavelength to 430 nm.

3. With the sample chamber empty, turn the left knob until the meter reads 0% transmittance (or A = ∞).

4. Fill a 13 × 100 mm test tube or cuvette with deionized water. Wipe it clean with a paper towel. Put it into the sample chamber and adjust the lower right knob until the meter reads 100% transmittance (A = 0.00).

5. Repeat numbers 3 and 4.

6. Remove the test tube containing deionized water from the sample chamber and replace it with a 13 × 100 mm test tube (wiped clean) containing solution #5. Read the absorption value from the **bottom** scale on the meter.

7. Repeat steps 2 through 6 (using only the water tube and test tube #5) at 460 nm, 480 nm, 500 nm, 510 nm, 520 nm, 540 nm, 570 nm and 610 nm.

8. Plot the absorption (y axis) versus the wavelength (x axis) and determine the wavelength of maximum absorption.

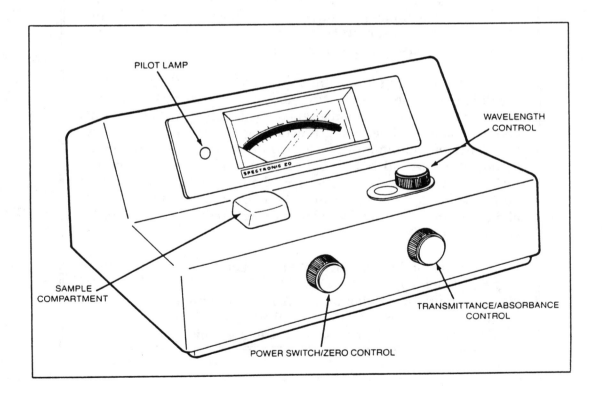

Figure 18-1
Reprinted with the courteous permission of the Milton Roy Company.

Chemical Capsule

The elements oxygen (65%), carbon (18%), hydrogen (10%), nitrogen (3%), calcium (1.5%), phosphorus (1.2%), potassium (0.2%), sulfur (0.2%), chlorine (0.2%), sodium (0.1%) and magnesium (0.05%) compose more than 99% of the human body. In addition, there are trace but **essential** amounts (<0.05%) of iron, cobalt, copper, zinc, iodine, selenium and fluorine. Aluminum, bromine, chromium, manganese, molybdenum and silicon are also found in the human body.

While small amounts of the trace elements are needed, the range of concentrations tolerated by the body are often narrow. This means that consumption of excessive supplements of the trace elements can cause more harm than good.

The topic of the spectroscopic analysis in this experiment, cobalt ion, is a case in point. Vitamin B_{12} is a coordination complex of cobalt and plays a role in the synthesis of nucleoproteins, proteins and red blood cells and the functioning of the nervous system. Deficiency of vitamin B_{12} leads to pernicious anemia. Only minute amounts are needed but its only source is from animal products including milk. Absolute vegetarians should supplement their diets with vitamin B_{12}. Cobalt has an LD_{50} on the order of 0.1 g/kg thus about 6 g of cobalt can be fatal and big excesses of vitamin B_{12} should be avoided.

In one of the most elegant and difficult accomplishments of organic chemistry, the late Dr. R. W. Woodward and a team of chemists synthesized vitamin B_{12} in 1973. For his very significant contributions to the field of synthetic organic chemistry, Dr. Woodward was awarded a Nobel prize in chemistry.

Prelaboratory Exercises - *Experiment 18*

For solutions to the starred problems, see *Appendix A.* For additional problem solving experience with solution concentrations, dilutions and the relationships between energy, frequency and wavelength, see *Exercises 9, 21.*

Name_____Date_____Lab Section_____

1.* Show the calculations that verify that 2.18 grams of $Co(NO_3)_2 \cdot 6H_2O$ dissolved in 50.00 mL of solution is a 0.150 M solution.

2. How many grams of copper sulfate pentahydrate need to be diluted to a total volume of 250.0 mL to make a 0.125 molar solution?

   _____

3. What is the concentration of a silver nitrate solution prepared by dilution of 0.500 g of silver nitrate to 2.000 L with water.

   _____

4. 3.00 mL of 6.0 M hydrochloric acid are diluted to 500.0 mL with water. What is the resulting concentration of hydrochloric acid?

   _____

5. How many mL of concentrated ammonia (14 M) should be diluted to 250 mL to make a 0.75 M NH_3 solution?

   _____

6.* 2.47×10^{-3} moles of sodium salicylate are diluted to 250 mL with water. 1, 2, 3, 4 and 5 mL aliquots of the solution are each diluted to 100 mL with excess iron(III) chloride solution to form a 1:1 iron salicylate complex. The absorption spectrum of these solutions reveals an absorption maximum at 530 nm. The absorption of each of the solutions (numbered according to the number of milliliters of the original solution diluted to 100 mL) at 530 nm in equal pathlength cells is given below. Calculate the concentration of iron salicylate in each solution and plot the concentration (x axis) versus the absorption (y axis). From the graph, determine the concentration of an iron salicylate solution of unknown concentration that gives an A value of 0.62.

solution (number of mL of stock solution diluted to 100 mL)	concentration [iron(III) salicylate complex (mole/L)]	absorption
1	_____	0.17
2	_____	0.33
3	_____	0.51
4	_____	0.68
5	_____	0.86
unknown	_____	0.62

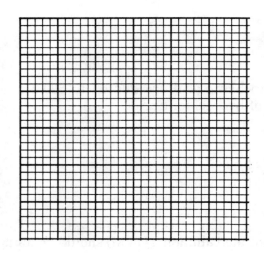

Results and Discussion - *Experiment 18*
SPECTROSCOPY

Name_____Date_____Lab Section_____

A. Preparation of solutions.

1. Formula mass of $Co(NO_3)_2 \cdot 6H_2O$ _____

2. Mass of beaker + $Co(NO_3)_2 \cdot 6H_2O$ _____

3. Mass of beaker   _____

4. Mass of $Co(NO_3)_2 \cdot 6H_2O$ (in the 50 mL volumetric flask)   _____

5. Moles of $Co(NO_3)_2 \cdot 6H_2O$ (in the 50 mL volumetric flask)   _____

6. Molarity of $Co(NO_3)_2$ solution   _____

B. Beer's law graph.

1. Unknown identification number   _____

2. Beer's Law Plot

Tube #	Concentration [calculated from A-6 and $M_1V_1 = M_2V_2$ (moles/L)]	Absorption
1	_____	_____
2	_____	_____
3	_____	_____
4	_____	_____
5	_____	_____
unknown	_____	_____

3. On the accompanying piece of graph paper, plot the absorption (y axis) versus the cobalt(II) nitrate concentration (x axis). From the graph and your measured absorption value for your unknown, determine the concentration of cobalt(II) nitrate in your unknown and fill in the the value in the table above.

4. Was Beer's Law obeyed? Explain your answer.   _____

C. Absorption spectrum of cobalt(II) nitrate.

Option 1 - the shorter, nonexperimental procedure

1. Absorption values were measured at several wavelengths for a 0.150 M $Co(NO_3)_2$ solution. Plot A (vertical axis) versus the wavelength (horizontal axis) and answer the questions below.

wavelength (nm)	absorption	wavelength (nm)	absorption
380	0.02	520	0.68
400	0.04	530	0.58
420	0.09	540	0.44
440	0.22	560	0.19
460	0.42	580	0.08
480	0.55	600	0.05
490	0.64	640	0.03
500	0.68	680	0.03
510	0.71		

2. According to your graph, what is the optimum wavelength for a Beer' law study?   _____

3. Calculate the percent difference between your measured value for the 0.150 M $Co(NO_3)_2$ solution at 510 nm and the value provided on this page. Suggest reasons for any difference between your value and the one measured by these authors.   _____

Option 2 - Experimental determination

1. Absorption versus wavelength for 0.150 M $Co(NO_3)_2$ (tube #5)

wavelength (nm)	absorption	wavelength (nm)	absorption
430	_____	520	_____
460	_____	540	_____
480	_____	570	_____
490	_____	610	_____
510	_____		

2. Graph A (y axis) vs λ (x axis) and determine the optimum wavelength. Explain your selection.   _____

Experiment 19

ACIDITY AND pH

Learning Objectives

Upon completion of this experiment, students should have learned:
1. How to evaluate acid strength.
2. The meaning of "pH".
3. How to determine the ionization constant for a weak acid.
4. The concepts of buffers and hydrolysis reactions.

Text Topics

Acids, bases, pH, buffers, ionization constants (Malone, Chapters 11, 12)

Comments

Predictions of pH values of the solutions for *Procedures A and B* should be made before the laboratory measurements are made. For additional problem solving experience with acidity and pH, see *Exercise 22.*

Discussion

The terms "acid" and "base" probably already have some meaning to you perhaps even invoking some fear in you. But acids and bases should be understood rather than feared as we encounter them everyday in the orange juice we drink, the vinegar we put on our salad, the baking soda we cook with and in many other household products. Chemists use specific descriptions of the chemical behavior of acids and bases to define them. We will use the Bronsted concept which states that an **acid** is a proton donor and a **base** is a proton acceptor. In aqueous solutions acids donate their protons to water to yield hydronium ions (or even more complex hydrated protons).

$$HCl(aq) \quad + \quad H_2O(l) \quad = \quad H_3O^+ \quad + \quad Cl^-$$

| Bronsted acid | Bronsted base | Conjugate acid | Conjugate base |

The above reaction is usually written in the simplified form:

$$HCl_{(aq)} \;=\; H^+ \;+\; Cl^-$$

where H^+ indicates the hydrated proton. Strong acids are almost completely ionized in water and significantly increase the hydrogen ion concentration of the solution. For example, 0.1 M HCl would have a 0.1 M hydrogen ion concentration as HCl ionizes almost completely. Weak acids only partially ionize and do not increase the hydrogen ion concentration nearly as much as equivalent amounts of a strong acid.

For equal concentrations of acids, it is possible to compare acid strength by measuring the hydrogen ion concentrations. Usually these measurements are performed in dilute solutions where the hydrogen ion concentrations are 10^{-1} to 10^{-14} M. To avoid using such small numbers, acidities are usually expressed logarithmically as pH values. The letter p means $-\log_{10}$. Thus pH means $-\log_{10}[H^+]$. As pH is the negative of the log, the lower the pH, the higher the acidity. For example, if the hydrogen ion concentration, $[H^+] = 1 \times 10^{-2}$ M, the pH $= 2$. If $[H^+] = 1 \times 10^{-5}$ M, the pH $= 5$. An instrument called a pH meter has been developed to accurately read out pH values directly. To determine the hydrogen ion concentrations, antilogarithms are taken. For example, if the pH of a solution is 3, $\log[H^+] = -3$ so $[H^+] = 10^{-3}$ M.

Base strength varies inversely with acid strength. The equilibrium expression for the dissociation of water,

$$H_2O_{(l)} \;=\; H^+ \;+\; OH^-$$

can be written as:

$$K_w \;=\; 10^{-14} \;=\; [H^+][OH^-] \qquad \textit{(equation 1)}$$

$$[OH^-] \;=\; 10^{-14}/[H^+] \qquad \textit{(equation 2)}$$

$$[H^+] \;=\; 10^{-14}/[OH^-] \qquad \textit{(equation 3)}$$

If logarithms are taken of both sides of *equation 1*, the following useful equation results:

$$14 \;=\; pH \;+\; pOH \qquad \textit{(equation 4)}$$

Equations 1 - 4 can be used to calculate the remaining three values of $[H^+]$, pH, pOH and $[OH^-]$ if one of the values is known or has been measured.

For a neutral solution, $[H^+] = [OH^-]$. Solving *equation 1* for this situation yields:

$$[H^+] \;=\; [OH^-] \;=\; 10^{-7}\ M \qquad \text{or} \qquad pH \;=\; pOH \;=\; 7.$$

Solutions with pH values less than 7 are acidic and solutions with pH values greater than 7 are basic. For two acid solutions of the same concentration, the solution with the lower pH is the stronger acid. Also note that a solution with a pH one unit lower than another is 10 times more acidic due to the nature of a logarithm scale.

Another way to compare acid strength involves the concept of the equilibrium constant. It is possible to show for the ionization of an acid:

$$HA_{(aq)} = H^+ + A^-$$

that the concentrations of products and reactants after equilibrium has been attained are related to an equilibrium constant by the equation:

$$K_a = \frac{[H^+][A^-]}{[HA]} \qquad \text{(equation 5)}$$

Notice from *equation 5* that the greater the percentage of ionization (or the stronger the acid) the larger the K_a value. K_a values are useful for comparison of acid strengths for weak acids. For strong acids, K_a values are very large (about 10 to 10^9). Your results today will enable you to make two, approximate determinations of the K_a value for acetic acid. See the *Prelaboratory Exercises* for the calculation procedure.

Some formulas such as Na_3PO_4 would not immediately make us think of acids or bases. Many compounds, however, undergo a hydrolysis reaction to some extent which causes the resulting solution to be acidic or basic. For example, sodium phosphate (Na_3PO_4 - sometimes called TSP for trisodium phosphate) ionizes in water. The phosphate ions have a tendency to react with water to produce a basic solution.

$$Na_3PO_{4(aq)} = 3 Na^+ + PO_4^{3-}$$

$$PO_4^{3-} + H_2O_{(l)} = HPO_4^{2-} + OH^-$$

Although the position of equilibrium may favor the unreacted phosphate ion, enough phosphate ions react with water enough to cause the resulting aqueous solution to have a pH value greater than 7.

Because of their occurrence in many biochemical systems including your own, buffers are another very important acid - base concept. Buffers are solutions made up of a weak acid and its conjugate base (such as acetic acid and sodium acetate) or a weak base and its conjugate acid (such as ammonia and ammonium chloride or carbonate and hydrogen carbonate). Buffers in your body prevent significant pH changes when an acid or base enters the system. In other words, buffers resist pH changes.

This experiment includes four parts. In the first part, your instructor will measure the pH values of several solutions (or at the instructor's option students can do their own measurements) and you will interpret observations in terms of acid or base strength, hydrolysis or buffer concepts. In the second part, you will use pH paper to estimate the pH of the same solutions measured with the pH meter. Third, you will compare the pH change in a buffer to a nonbuffer when a base is added. Finally you will be given four dropping bottles containing solutions of acids and bases at various concentrations. Your problem will be to group the solutions into two categories, acids and bases, and to rank each group according to increasing concentration.

Procedure

A. pH measurement using a pH meter. Before any measurements are made, predict the pH values for each of the solutions listed in the *Results and Discussion* section (some of the solutions e.g., #1, 2, 3, 13 can be predicted quantitatively while the others can only be estimated to within 1 or 2 pH units). Then you or your instructor will measure the pH values of the solutions.

B. pH measurement using pH paper. Using broad range pH paper (1 - 11), determine the pH of the solutions listed in the *Results and Discussion* section and/or any other solutions provided by the instructor. To conserve pH paper, tear off a small piece and put a drop of the solution on the paper rather than dipping a long strip into the solution.

C. Buffers. Two solutions, *A* and *B*, will be provided.

$$A = 1 \times 10^{-3} \text{ M HCl} \qquad\qquad B = 0.1 \text{ M HC}_2\text{H}_3\text{O}_2 + 0.1 \text{ M NaC}_2\text{H}_3\text{O}_2$$

Transfer about 25 mL of solution *A* to a small beaker, insert the calibrated pH electrode and read the pH (or use pH paper to determine the pH). Leave the pH electrode in the solution, add 1 drop of 1 M NaOH, stir and read the pH (or use pH paper again). Add 9 more drops of 1 M NaOH, stir and read again (or use pH paper again). Add 15 more drops and repeat the reading (or use pH paper again). Clean out the beaker and repeat the experiment with solution *B*.

D. Unknown acids and bases. You will be provided with four bottles labeled *1 - 4* that contain the four solutions below:

0.1 M NaOH	0.1 M HCl
0.4 M NaOH	0.4 M HCl

The order of the solutions will not correspond to the above list. The acids will all contain a small amount of phenolphthalein, an indicator that is colorless in acid and pinkish-red in base. By determining when a combination of two solutions results in a neutralization reaction (a color change due to phenolphthalein) and the number of drops of each solution used for the neutralization, you should be able to group the solutions into acids and bases and rank each group according to increasing acid or base strength. Be sure to consider the order of mixing when attempting to predict observations.

Prelaboratory Exercises - *Experiment 19*

For solutions to the starred problems, see *Appendix A*. For additional problem solving experience with acidity and pH, see *Exercise 22*.

Name_____Date_____Lab Section_____

1.* Fill in the blanks in the following table:

pH	[H⁺] (moles/L)	pOH	[OH⁻] (moles/L)
4	_____	_____	_____
_____	1×10^{-6}	_____	_____
_____	_____	_____	1×10^{-3}
_____	_____	5	_____
1.44	_____	_____	_____
_____	7.2×10^{-9}	_____	_____

2. Fill in the blanks in the following table:

pH	[H⁺] (moles/L)	pOH	[OH⁻] (moles/L)
9	_____	_____	_____
_____	1×10^{-4}	_____	_____
_____	_____	_____	1×10^{-6}
_____	_____	7.50	_____
3.73	_____	_____	_____
_____	4.6×10^{-11}	_____	_____

3.* What is the expected pH of a 1.0×10^{-2} M HNO_3 solution?   _____

4. How many times more acidic is a solution with pH = 2 than one with pH = 5? _____

5.* A 1.0×10^{-1} M H_2S solution has a pH of about 4. What are the concentrations of H^+, SH^-, and H_2S in the solution and what is K_a for the extremely toxic gas, H_2S?

6.* An aqueous solution containing 0.1 M HF and 0.1 M KF has a pH of 3.45. What is the K_a for HF?

7.* A solution of sodium acetate has a pH of about 9. Write a net ionic equation that accounts for this pH.

8. For *Part A*, fill in the prediction column in the table.

9. Prepare a prediction matrix for *Part D*. Be sure to fill in expected observations for both directions of mixing.

Results and Discussion - *Experiment 19*
ACIDITY AND pH

Name_____Date_____Lab Section_____

A, B. pH measurements.

#	Solution	pH (predicted)	pH (paper)	pH (electronic)
1.	0.1 M HCl	_____	_____	_____
2.	0.01 M HCl	_____	_____	_____
3.	0.001 M HCl	_____	_____	_____
4.	0.1 M $HC_2H_3O_2$	_____	_____	_____
5.	0.1 M NH_4Cl	_____	_____	_____
6.	deionized water	_____	_____	_____
7.	tap water	_____	_____	_____
8.	0.1 M NaCl	_____	_____	_____
9.	0.1 M $NaC_2H_3O_2$	_____	_____	_____
10.	0.1 M $NaHCO_3$	_____	_____	_____
11.	0.1 M Na_2CO_3	_____	_____	_____
12.	0.1 M NH_3	_____	_____	_____
13.	0.1 M NaOH	_____	_____	_____
14.	vinegar	_____	_____	_____
15.	orange juice	_____	_____	_____
16.	Seven-Up	_____	_____	_____
17.	milk of magnesia	_____	_____	_____
18.	milk	_____	_____	_____

19. Using the pH values determined for solutions *1, 4, 12, 13*, calculate $[H^+]$, pOH, $[OH^-]$.

solution #	pH	$[H^+]$ (moles/L)	pOH	$[OH^-]$ (moles/L)
1	_____	_____	_____	_____
4	_____	_____	_____	_____
12	_____	_____	_____	_____
13	_____	_____	_____	_____

20. Compare and explain the pH differences between solutions *1* and *4*.

21. Use the concentrations and pH values for *solution 4* to calculate an approximate K_a for acetic acid. _____

$$HC_2H_3O_{2(aq)} = H^+ + C_2H_3O_2^-$$

$$K_a = \frac{[H^+][C_2H_3O_2^-]}{[HC_2H_3O_2]}$$

22. Write net ionic equations that account for the pH values observed in numbers 5, 11 and 12.

#5

#11

#12

23. Assuming that the pH meter values are correct, comment on the reliability of pH paper for measuring pH values.

C. Buffers

A = 1x10⁻³ M HCl

B = 0.1 M $HC_2H_3O_2$ + 0.1 M $NaC_2H_3O_2$

total drops of 1 M NaOH	pH	total drops of 1 M NaOH	pH
0	_____	0	_____
1	_____	1	_____
10	_____	10	_____
25	_____	25	_____

1. Compare your observations for the addition of 1 M NaOH to solutions A and B and account for any differences.

2. Use the concentrations and pH value for the first pH measurement of solution B to calculate an approximate K_a for acetic acid.   _____

D. Unknown acids and bases.

1. For each solution listed, give the code from the bottle.

bases	code	acids	code
0.1 M NaOH	_____	0.1 M HCl	_____
0.4 M NaOH	_____	0.4 M HCl	_____

2. Briefly describe the significant results of your experiments with bottles 1 - 4 that led you to the above assignments.

3. Suggest any ways you can think of to improve any part(s) of this experiment.

Experiment 20

SYNTHESIS

Learning Objectives

Upon completion of this experiment, students should have learned:
1. Synthetic methods used in chemistry.
2. To determine the percent yield of a product.

Text Topics

Isomers, organic synthesis (Malone, Chapters 7, 16)

Comments

The products of this experiment are used in later experiments (*21 and* 23) and should be saved. For problem solving experience with classification of organic functional groups, see *Exercise 27.*

Discussion

One of the principal goals of the chemical industry is the preparation of new and better chemicals for medicines, pesticides, fertilizers, clothes, food additives and virtually every other household item. The chemical preparation process is called synthesis and is the subject of today's experiment which includes two rather short syntheses. The first is the synthesis of the copper - amino acid complex, copper(II) glycinate monohydrate. The second is a synthesis of the natural product benzoic acid, a compound often used as a food additive.

A chemical synthesis is usually followed by proper purification and identification techniques. Next week, you will titrate the synthesized acid to determine its molecular mass. In *Experiment 23*, you will determine the percent copper in the copper glycinate complex.

As economics is a prime concern of the chemical industry, the percent yield of the reaction is a very important consideration. The percent yield is calculated by dividing the amount of product experimentally obtained by the theoretical amount (amount calculated from amounts of starting material used) and multiplying by 100%.

A. Copper(II) glycinate monohydrate. One morning in 1892, French-born Swiss chemist Alfred Werner awoke from a dream with the solution to problems about molecular structure that were perplexing both himself and his contemporaries. This was the birth of coordination chemistry. This rather strange technique of solving difficult chemical problems was not unprecedented. Some years earlier, clues to the structure of benzene and spatial aspects of carbon bonding had come to August Kekulé in dreams. The concept that Werner had dreamt was that certain compounds consist of a central metal ion surrounded by a definite number of atoms. Often the atoms are part of other molecules such as ammonia or water. In many cases, two atoms of the same molecule will coordinate around the metal ion. These compounds are called chelates. The term Werner chose for the number of atoms that surround the central ion is "coordination number." Many coordination compounds have *"cis"* and *"trans"* isomers. Atoms of the same kind are on the same side of the molecule in the *cis* configuration but on opposite sides in the *trans* configuration. For example, in today's experiment, you will be synthesizing the chelate *cis* copper(II) glycinate monohydrate. If the *cis* isomer is heated to 180°C for a few minutes, it converts to the *trans* isomer.

```
O = C —— CH₂            O = C —— CH₂
    |       |               |       |
    O      NH₂              O      NH₂
     \    ,                  \    ,
       Cu                      Cu
     /    ,                  ,    \
    O      NH₂             NH₂     O
    |       |               |       |
O = C —— CH₂             H₂C —— C = O
```

cis-copper(II) glycinate   trans-copper(II) glycinate

Werner's concept of coordination complexes was so radically different from accepted chemical concepts at the time that it took a number of years and countless experiments to convince the skeptics. The validity of Werner's theory has now been thoroughly demonstrated. Coordination complexes are distributed widely in nature. For example, heme, which is necessary for the transportation of oxygen in blood, is a coordination complex of iron(II). Another well known complex, chlorophyll, is a coordination complex of magnesium(II). Chlorophyll is the green substance in leaves that is necessary for the photosynthetic reactions that convert carbon dioxide and water to glucose and oxygen. Copper(II) is one of the ions that forms coordination complexes and a number of its complexes are found in nature. Although copper(II) ion is toxic in appreciable amounts, it is necessary as a trace element in human nutrition. It is instrumental in the function of enzymes that are involved in mitochondrial energy production and formation of melanin, elastin, bone and nervous tissue. Copper(II) ion is absorbed in the stomach and upper small intestine by forming complexes with amino acids and peptides. *Cis*-copper(II) glycinate monohydrate is one of these copper - amino acid complexes.

Today's preparation of copper(II) glycinate monohydrate can be viewed as a double replacement reaction between copper(II) acetate and glycine:

$$Cu(C_2H_3O_2)_{2(aq)} + 2\ HC_2H_4NO_{2(aq)} = Cu(C_2H_4NO_2)_{2(s)} + 2\ HC_2H_3O_{2(aq)}$$

Glycine and acetic acid are actually similar in structure:

$$\underset{\underset{\displaystyle NH_2 \;\; glycine}{|}}{CH_2}-\overset{\displaystyle \overset{O}{\|}}{C}-OH \qquad\qquad\qquad CH_3-\overset{\displaystyle \overset{O}{\|}}{C}-OH$$

acetic acid

Glycine is one of the 20 amino acids your biological processes join together with other amino acids to form peptides and proteins.

B. Benzoic acid. Benzoic acid is a naturally occurring organic acid that is present in large quantities in cherries and most berries. Because it has been found to have fungicidal properties, it is often used as a food preservative. In this experiment, you will prepare benzoic acid from methyl benzoate via a reaction known as a saponification reaction. The reaction consists of two steps, saponification of methyl benzoate:

methyl benzoate	+	sodium hydroxide	$\longrightarrow$	sodium benzoate	+	methanol
$C_8H_8O_2$		NaOH		$C_7H_5O_2Na$		CH_3OH

and the reaction of sodium benzoate with acid to give benzoic acid:

sodium benzoate	+	hydrochloric acid	$\longrightarrow$	benzoic acid	+	sodium chloride
$C_7H_5O_2Na$		HCl		$C_7H_6O_2$		NaCl

Procedure

A. **Copper(II) glycinate monohydrate.** Weigh about 1.6 grams of copper(II) acetate monohydrate to at least the nearest 0.01 g into a 250 mL beaker and add 15 mL of deionized water. Weigh out about 1.3 g of glycine into a 150 mL beaker and add 10 mL of deionized water. Place both beakers on a hot plate and heat to the boiling point. Stir occasionally while heating until both solids dissolve. After the solids dissolve and the solutions are near the boiling point, **carefully** remove the beakers (use beaker tongs) from the hot plate and add the glycine solution to the copper(II) acetate solution and stir. Allow the solution to cool for several minutes and place the beaker in an ice bath. After crystals begin to form, add 25 mL of 1-propanol with continuous stirring. Continue to cool for several minutes and vacuum filter (use a Buchner funnel and a filter flask - see *Figure 20-1*) to collect the product. Wet the filter paper first to insure a good seal. Turn the water on full for the aspirator to operate efficiently. Rinse the beaker to make sure all solid is transferred to the funnel with acetone from a wash bottle (**keep away from flames**) and add the wash liquid to the funnel. Wash the precipitate with acetone twice and continue to pull air through the sample for a few minutes. Then transfer the solid to a weighed piece of filter paper. Allow the sample to dry for a few days and then weigh it to determine the experimental and percent yields. Save the sample for analysis in *Experiment 23*.

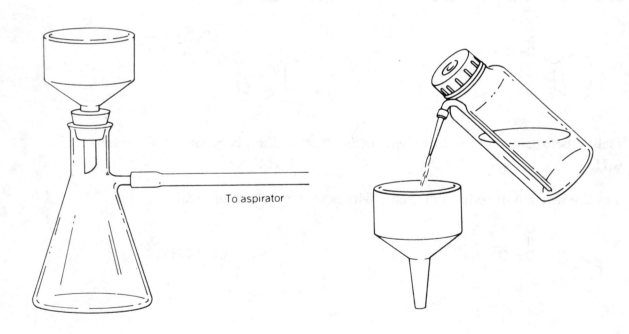

To aspirator

Figure 20-1

B. Benzoic acid. Heat a 600 mL beaker half full of deionized water to boiling on a hot plate. Put 3 mL of methyl benzoate, 8 mL of 1-propanol and 10 mL of 6 M sodium hydroxide into a 125 mL Erlenmeyer flask. Swirl the contents and place the flask in the beaker of boiling water for 15 minutes (*Figure 20-2*). Remove the flask from the water bath and place it in an ice bath for a few minutes until it is cool. If any precipitate forms at this point, break it up with a stirring rod or spatula. Slowly and cautiously add 12 mL of 6 M HCl (*CAUTION: RAPID ADDITION COULD CAUSE SOME OF THE CONTENTS OF THE FLASK TO BOIL OUT OF THE FLASK IF THE MIXTURE HASN'T BEEN SUFFICIENTLY COOLED FIRST*) to the flask in the ice bath. Continue to cool the flask in the ice bath for 10 minutes breaking up large lumps of precipitate that form with a spatula or stirring rod. Collect the precipitate with a Buchner funnel and filter flask as in *Procedure A* above. Wash the precipitate three times with ice-cold water from a wash bottle. Continue to pull air through the precipitate for several minutes after the third washing. Spread the sample on a piece of filter paper and allow it to dry for a few days and weigh it to determine the percent yield.

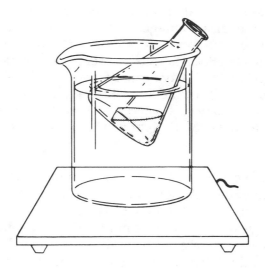

Figure 20-2

Chemical Capsule

The next time you have the opportunity, ask some elementary school children if they use any synthetic (made by humans) chemicals. You will probably be surprised to find out that many cannot think of any. When they finally do, it is usually chemicals like make-up or hair spray. If you probe further, the children begin to realize that they are surrounded by synthetic chemicals. Eventually the students usually name cleaning agents, plastics, explosives, clothing materials, food additives, pesticides, computer chips, gasoline and medicines. In fact, they then realize that it is hard to name "natural" chemicals that they use.

Despite the fact that the advantages of synthetic chemicals far outweigh the disadvantages, many people focus only on the problems caused by chemicals and never notice how dependent they are on the beneficial aspects of chemicals. While it is true that pesticides such as DDT have caused considerable damage to our ecology, people don't realize that DDT saved millions of lives. We should, however, make sure a chemical is safe before using it rather that waiting to determine after using it that it has detrimental results.

Before World War II, it was common for bacterial infections to be life threatening. Although Fleming discovered that penicillin kills bacteria in 1928, it took 15 years before scientists finally developed ways of effectively administering penicillin. Since then we have lived in an era with bacterial diseases basically under control. Recently, however, it has been found that some bacteria have mutated to the point that very few antibiotics are effective against them. We need to very quickly develop new antibiotics to combat this threat or we may enter a new era where bacteria commonly kill once again.

Prelaboratory Exercises - *Experiment 20*

For solutions to the starred problems, see *Appendix A*.

Name_____Date_____Lab Section_____

It is possible to use a two step reaction sequence to synthesize aspirin from oil of wintergreen (methyl salicylate).

1.* In the first step, 25 grams of methyl salicylate was reacted with an excess of base followed by acidification and 16 grams of salicylic acid was obtained. Calculate the theoretical and percent yields of salicylic acid.

Overall reaction: _____

$C_8H_8O_3$ + H_2O = $C_7H_6O_3$ + CH_4O _____
methyl   salicylic   methanol
salicylate   acid

2.* In the second step, the 16 grams of salicylic acid was reacted with an excess of acetic anhydride and 16 grams of aspirin (acetylsalicylic acid) was obtained. Calculate the theoretical and percent yields of aspirin.

$C_7H_6O_3$ + $C_4H_6O_3$ = $C_9H_8O_4$ + $C_2H_4O_2$ _____
salicylic   acetic   aspirin   acetic
acid   anhydride   acid   _____

3. 7.0 g of barium hydroxide octahydrate and 7.0 g of ammonium thiocyanate are mixed together in the solid state and the following reaction occurs:

___$Ba(OH)_2 \cdot 8H_2O(s)$ + ___$NH_4SCN(s)$ = ___$Ba(SCN)_2(aq)$ + ___$NH_3(aq)$ + ___$H_2O(l)$

Balance the equation, calculate the amount of barium thiocyanate that should theoretically be formed and the amount of the reagent present in excess that is actually consumed (or needed) for the reaction.

_____

_____

4. 3.0 g of cyclohexene (C_6H_{10}) reacts with 9.0 g of bromine to give 5.0 g of dibromocyclohexane ($C_6H_{10}Br_2$).

$$C_6H_{10}(l) \quad + \quad Br_2(l) \quad = \quad C_6H_{10}Br_2(l)$$

Determine the theoretical and percent yields of dibromocyclohexane and the limiting reagent in the reaction.

    _____

    _____

5. One method of synthesizing barium sulfate would be to mix aqueous solutions of barium chloride and sodium sulfate. The resulting double replacement reaction yields insoluble barium sulfate which could be collected by filtration. Using *Appendix C* as a guide, suggest solutions that could be mixed together to make the following products (Remember that the reactants must be soluble in water).

    a. calcium carbonate

    b. zinc oxalate

    c. silver chromate

Results and Discussion - *Experiment 20*
SYNTHESIS

Name_____Date_____Lab Section_____

A. Copper(II) glycinate monohydrate.

$$Cu(C_2H_3O_2)_2 \cdot H_2O + 2\, HC_2H_4NO_2 = Cu(C_2H_4NO_2)_2 \cdot H_2O + 2\, HC_2H_3O_2$$

1. Mass of copper(II) acetate monohydrate _____

2. Formula mass of copper(II) acetate monohydrate _____

3. Moles of copper(II) acetate monohydrate _____

4. Mass of glycine _____

5. Formula mass of glycine _____

6. Moles of glycine _____

7. Formula mass of copper(II) glycinate monohydrate _____

8. Theoretical yield of copper(II)glycinate monohydrate based on copper(II) acetate monohydrate (show series of unit conversions) _____

9. Theoretical yield of copper(II)glycinate monohydrate based on glycine (show series of unit conversions) _____

10. Limiting reagent _____

11. Mass of filter paper _____

12. Mass of filter paper + product _____

13. Mass of copper(II) glycinate monohydrate _____

14. Percent yield of copper(II) glycinate monohydrate _____

B. Benzoic acid

overall reaction:

$$C_8H_8O_2 + NaOH + HCl = HC_7H_5O_2 + CH_3OH + NaCl$$
methyl   benzoic
benzoate   acid

1. Milliliters of methyl benzoate   _____

2. Mass of methyl benzoate (density of methyl benzoate = 1.09 g/mL)   _____

3. Molecular mass of methyl benzoate   _____

4. Moles of methyl benzoate   _____

5. Theoretical yield in moles of benzoic acid (assume that methyl benzoate was the limiting reagent)   _____

6. Molecular mass of benzoic acid   _____

7. Theoretical yield of benzoic acid in grams   _____

8. Experimental mass of benzoic acid obtained   _____

9. Percent yield of benzoic acid   _____

10. Show the series of unit conversions you could use to calculate the theoretical yield of benzoic acid in grams from the volume, density and molecular mass of methyl benzoate, the molar ratio and molecular mass of benzoic acid.

11. (Optional) Melting range of synthesized benzoic acid (literature value - 122°C)   _____

12. Suggest any ways you can think of to improve any part(s) of this experiment.

Experiment 21

ACID-BASE TITRATIONS

Learning Objectives

Upon completion of this experiment, students should have learned:
1. To determine the concentration of an acid using titration.
2. To determine the formula mass of a compound using titration.
3. Skills of titration techniques.

Text Topics

Acid-base reactions and titrations, solution stoichiometry (Malone, Chapter 11, 12)

Comments

An experimental determination of the molecular mass of the benzoic acid prepared in *Experiment 20* will be performed. For additional problem solving experience involving solution stoichiometry, see *Exercise 21*.

Discussion

In chemical analysis, it is often necessary to determine the concentration of ions in solution. For this purpose, a technique called titration (recall the water hardness determination) is often utilized. In a titration, a solution is added using a buret to a known volume or mass of another reactant until the reaction is stoichiometrically complete. Completion of the reaction is detected by using an indicator or some type of instrumentation. From the results, it is possible to calculate a previously unknown quantity such as the concentration or molecular mass of one of the reactants.

Ideally suited for the titration method are determinations of acid or base concentrations or formula masses. There are many available indicators that dramatically change color at the stoichiometric end point of acid-base reactions.

$$HA \quad + \quad MOH \quad = \quad MA \quad + \quad H_2O$$
acid   base   salt   water

For most strong acid - strong base reactions the net ionic equation is:

$$H^+ \quad + \quad OH^- \quad = \quad H_2O$$

For the typical acid-base titration, an acid or base whose concentration is known to about 4 significant figures is required. In today's experiment, you will need a sodium hydroxide solution with an accurately known concentration. However, NaOH is impure and hygroscopic and the concentration of solutions prepared by diluting weighed amounts to known volumes cannot be accurately calculated. In order to find the concentration of an NaOH solution accurately, it is necessary to titrate it against a primary standard. Primary standards are compounds that are at least 99.95% pure and are not hygroscopic. Your instructor will provide you with NaOH that has been standardized by titrating a primary standard acid.

You will use the standardized NaOH to determine the concentration of an HCl solution and the molecular mass of benzoic acid. For calculation purposes, the molarity of the NaOH times the volume needed to reach the end point equals the moles of NaOH. The moles of acid are related to the moles of base by the molar ratio of acid to base as given by the ratio of coefficients from the balanced equation. While this ratio is often unity, assuming unity can often lead to errors. For diprotic acids such as H_2SO_4 or a base such as $Ba(OH)_2$, the molar ratio is generally not one.

Examples:

$$NaOH_{(aq)} \quad + \quad HCl_{(aq)} \quad = \quad NaCl_{(aq)} \quad + \quad H_2O_{(l)} \qquad \text{Molar ratio of acid to base is 1:1}$$

$$2\,NaOH_{(aq)} \quad + \quad H_2SO_{4(aq)} \quad = \quad Na_2SO_{4(aq)} \quad + 2\,H_2O_{(l)} \qquad \text{Molar ratio of acid to base is 1:2}$$

The moles of acid can be used to calculate molarity by dividing the moles of acid by the volume of acid used. Alternatively, the molecular mass of an acid can be determined by dividing the mass of the titrated sample by the number of moles of acid determined from the titration.

Procedure

A. Determination of the concentration of an acid. Transfer about 100 mL of a NaOH solution (~ 0.1 M) of known concentration into a **clean, dry** flask and stopper it. Check out a pair of burets (or a buret and a pipet) and obtain from your instructor a sample of HCl of unknown concentration. Wash the burets with detergent and rinse them with deionized water. Rinse the buret you will be putting your standard base into three times with a few milliliters of your standard base and discard the rinse each time. Make sure that the base solution coats the entire inside of the buret each time you rinse it. After three rinsings, fill your buret with base. Open the stopcock or pinchcock and shake the buret to remove any air that may be trapped in the tip. If you are using a pinchcock-type buret, a good way to remove air is to aim the tip upward and to allow a little liquid to flow through. Air being less dense that the liquid should flow out before the liquid does. Clamp the buret in a buret clamp.

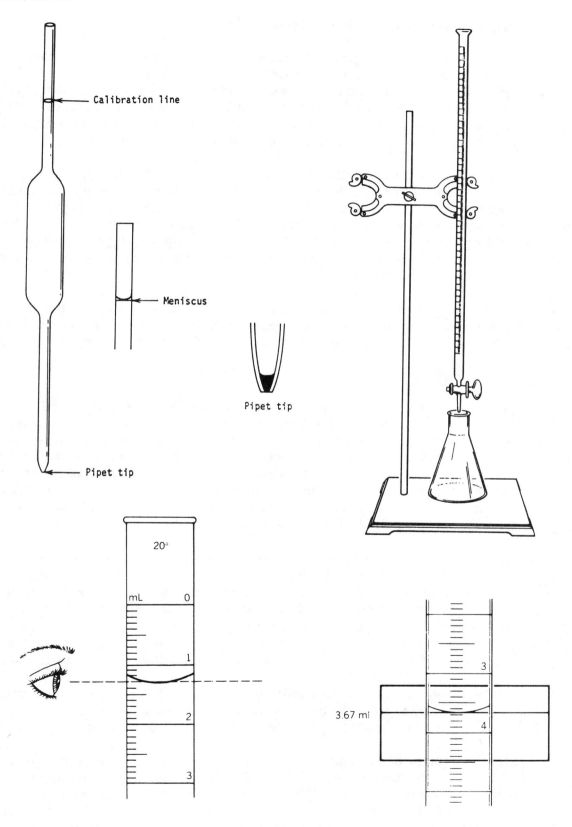

Calibration line

Meniscus

Pipet tip

Pipet tip

20°

mL   0

1

2

3

3.67 ml

3

4

Figure 21-1

Option 1 - Pipet for the acid, buret for the standardized base.

Rinse a 10.00 mL volumetric pipet three times with your unknown HCl solution. Deliver 10.00 mL of the acid into a clean (but not necessarily dry) 125 or 250 mL Erlenmeyer flask. Add two drops of phenolphthalein solution to the flask containing the unknown acid. Read the buret containing the standardized base to the nearest 0.01 mL. This means you must estimate how many tenths of the distance between the 0.1 mL lines the meniscus is below the tenth milliliter line above it. Take your reading at the bottom of the meniscus of the liquid.

Begin to run your base solution into your flask containing the acid solution and phenolphthalein. A pink color will develop which quickly disappears when the solution is swirled. As more base is added, the pink color will remain longer. When the pink color becomes spread out and remains for longer periods of time, slow the addition of your base so that you are <u>adding one drop or even ½ drop (by quickly rotating the stopcock 180°) at a time</u>. Swirl your flask with each addition. At the endpoint, one drop of base solution will turn the entire contents of the flask a light pink. At this point, record the final volume for the base.

Perform two more titrations of your unknown acid in the same way you performed the first. It won't be necessary to rinse the buret or pipet with the solutions contained in them this time because their walls are already coated with those solutions from the previous titration.

Option 2 - Burets for both the acid and base.

Rinse the second buret three times with your unknown HCl acid solution. Fill this buret with the acid solution and clamp it. Read the buret containing your unknown acid to the nearest 0.01 mL. This means you must estimate how many tenths of the distance between the 0.1 mL lines the meniscus is below the tenth milliliter line above. Take your reading at the bottom of the meniscus of the liquid. Now run about 15 mL of your unknown acid solution into a clean (but not necessarily dry) 125 or 250 mL Erlenmeyer flask and take a second reading. The difference between the two readings equals the volume of unknown acid introduced into the flask. Add two drops of phenolphthalein to the flask containing the unknown acid.

Now read the buret containing the base. Begin to run your base solution into your flask containing the acid solution and phenolphthalein. A pink color will develop which quickly disappears when the solution is swirled. As more base is added, the pink color will remain longer. When the pink color becomes spread out and remains for longer periods of time, slow the addition of your base so that you are adding it <u>one drop or even ½ drop</u> at a time. Swirl your flask with the addition of each drop. At the endpoint, one drop of base solution will turn the entire contents of the flask a light pink. At this point, record the final volume for the base.

Perform two more titrations of your unknown acid in the same way you performed the first. It won't be necessary to rinse the burets with the solutions contained in them this time because their walls are already coated with those solutions from the previous titration.

B. Molecular mass of an acid. The benzoic acid synthesized in *Experiment 20* is recommended for use as the "unknown acid" although other unknowns may be supplied by the instructor. Add 0.3 grams of the acid (weigh to at least the nearest 0.01 gram) to two or three Erlenmeyer flasks. Add about 5 mL of ethanol to each flask and swirl until the acid dissolves. Add three drops of phenolphthalein indicator and titrate to the end point. From your experimental data, determine the molecular mass of your acid.

Prelaboratory Exercises - *Experiment 21*

For solutions to the starred problems, see *Appendix A*. For additional problem solving experience involving solution stoichiometry, *Exercise 21*.

Name_____Date_____Lab Section_____

1.* 0.6530 grams of a primary standard acid (potassium hydrogen phthalate, molecular mass = 204.23 g/mole) require 23.32 mL of a NaOH solution of unknown concentration to reach the endpoint. What is the concentration of NaOH?

_____

2.* If 16.0 mL of a 0.120 M NaOH solution are required to neutralize 25.0 mL of an HCl solution, what is the concentration of the HCl solution?

_____

3.* If 21.3 mL of a 0.120 M NaOH solution are required to neutralize 0.400 grams of an unknown monoprotic acid, what is the molecular mass of the acid?

_____

4. If 14.0 mL of a 0.160 M NaOH solution are required to neutralize 11.0 mL of a sulfuric acid solution, what is the concentration of the H_2SO_4 solution?

_____

5. 0.3000 g of a diprotic acid requires 32.86 mL of 0.1100 M NaOH to reach the second end point. What is the molecular mass of the acid.

_____

6. What is the meaning of the terminology "primary standard" and why can't sodium hydroxide be used as a primary standard?

7.* When titrating hydrochloric acid with sodium hydroxide, at what pH should the indicator change color?

8. When titrating acetic acid with sodium hydroxide, should the indicator change color at:

a. a slightly basic pH
b. a pH of 7
c. a slightly acidic pH

Results and Discussion - *Experiment 21*
ACID-BASE TITRATION

Name_____Date_____Lab Section_____

A. Determination of the concentration of an acid.

1. Unknown number   _____

Volume of acid	#1	titration #2	#3
2. Final buret reading	_____	_____	_____
3. Initial buret reading	_____	_____	_____
4. Volume of acid[1]	_____	_____	_____

Volume of base

	#1	#2	#3
5. Final buret reading	_____	_____	_____
6. Initial buret reading	_____	_____	_____
7. Volume of base	_____	_____	_____

Calculations

	#1	#2	#3
8. Molarity of base	_____	_____	_____
9. Moles of base	_____	_____	_____
10. Moles of acid[2]	_____	_____	_____
11. Molarity of acid	_____	_____	_____

12. Average molarity of acid   _____

13. For titration *1*, show the series of unit conversions you could use to calculate the molarity of the acid from the volume and molarity of the NaOH solution, the acid to base mole ratio and the volume of the acid.

---

[1] A buret or pipet can be used to measure the amount of acid. If a pipet is used, record the volume directly on the third line.

[2] Assume that the acid is monoprotic (general formula HX). Therefore the number of moles of acid and base are equal in this case.

B. Molecular mass of an acid.

1. Sample identity (benzoic acid or unknown #) _____

Titration	#1	#2	#3
2. Mass of flask + acid	_____	_____	_____
3. Mass of flask	_____	_____	_____
4. Mass of acid	_____	_____	_____
5. Final buret reading	_____	_____	_____
6. Initial buret reading	_____	_____	_____
7. Volume of base	_____	_____	_____
8. Molarity of NaOH	_____	_____	_____
9. Moles of NaOH	_____	_____	_____
10. Moles of acid[3]	_____	_____	_____
11. Experimental molecular mass of acid	_____	_____	_____

12. Average experimental molecular mass of acid _____

13. Molecular mass of benzoic acid ($C_7H_6O_2$) from formula _____

14. Percent error of your experimental benzoic acid molecular mass _____

15. Does your experimental result help confirm that benzoic acid was the product of last week's synthesis? Explain your answer.

16. Suggest any ways you can think of to improve any part(s) of this experiment.

---

[3]Benzoic acid is monoprotic, thus the moles of acid equal the number of moles of base.

Experiment 22

OXIDATION-REDUCTION

Learning Objectives

Upon completion of this experiment, students should have learned:
1. The nature of oxidation-reduction reactions.
2. To make an activity series from experimental data.
3. To write half reactions.

Text Topics

Oxidation-reduction reactions, activity series (Malone, Chapter 13)

Comments

This is a relatively short experiment and can be performed with another short experiment during the same laboratory period. For additional problem solving experience with oxidation numbers, half reactions and balancing oxidation-reduction reactions, see *Exercises 23, 24.*

Discussion

In *Experiment 9*, we learned how to classify reactions according to the categories combination, decomposition, combustion, single replacement and double replacement. A second classification scheme differentiates between reactions in which electrons are transferred and those in which electrons are not transferred. If electrons are transferred from one reactant to another the reaction is called an oxidation-reduction or redox reaction. One method of checking for electron transfer is to compare oxidation numbers of the elements on the reactant side to the elements on the product side. If the oxidation number of any element has increased (e.g., $0 \rightarrow 2$, $2 \rightarrow 4$, $-1 \rightarrow 0$), the element has lost electrons and has been oxidized. As a result, another element must have gained electrons and have been reduced or experienced a decrease in oxidation number (e.g. $2 \rightarrow 0$, $0 \rightarrow -1$). Many reactions are easily detectable as redox reactions by the presence of a reactant or product in elemental form. An element cannot react or be produced unless electrons are transferred. This means that all single replacement reactions are also redox reactions. Combustion reactions are also redox reactions. Some combination and decomposition reactions are redox reactions but double replacements are never redox reactions. This experiment focuses on what is probably the simplest type of redox reaction, the single replacement reaction.

If the displaced element gains electrons in the reaction, it is **reduced** from its ionic form to its elemental form. The reactive metal or ion that supplied the electrons is simultaneously **oxidized**. Oxidation is the loss of electrons.

$$Zn° + Pb(NO_3)_{2(aq)} = Pb° + Zn(NO_3)_{2(aq)}$$

In the above equation, lead is reduced to its elemental form. Zinc is oxidized from its elemental form to its ionic form. This demonstrates that zinc metal is a more active reducing agent than lead metal. If no reaction occurs when a metal or reducing agent is introduced into a solution, the elemental form of the metal already in solution is more active or a better reducing agent than the metal that was put into the solution.

If the displaced element loses electrons in the reaction, it is oxidized from its ionic form to its elemental form. The reactive element or ion that acquires the electrons is reduced.

$$F_2 + 2 Br^- = 2 F^- + Br_2$$

Fluorine is reduced in the above equation while bromine is oxidized from its ionic form to its elemental form. This demonstrates that fluorine is a more active oxidizing agent than bromine. If no reaction occurs when an oxidizing agent is introduced into a solution, then the oxidized form of the negative ion already in solution is more active, or a better oxidizing agent than the oxidizing agent that was put into it.

In the first part of the experiment you will determine the relative reducing activities of Cu, Fe, H, Mg and Zn. Remember, if the elemental form of metal A replaces the ionic form of metal B in solution, metal A is a better reducing agent than metal B. If the elemental form of metal A does not replace the ionic form of metal B, then metal B is a better reducing agent than metal A. [Note: Magnesium metal reacts with water slowly to give magnesium hydroxide and hydrogen gas. When investigating the reactivity of elemental magnesium, look for evidence of a reaction other than the formation of bubbles on the surface of the metal.]

In the second part of the experiment you will determine the relative oxidizing activities of Br_2, Cl_2, I_2 and Fe^{3+}. Remember, if Fe^{3+} or the elemental form of a halogen [group VIIA (or 17) elements are called halogens] replaces the ionic form of another halogen, the first halogen (or Fe^{3+}) is a better oxidizing agent than the other halogen. If the elemental form of a halogen (or Fe^{3+}) does not replace the ionic form of the other halogen, then the halogen in ionic form is the better oxidizing agent.

HALF REACTIONS

Complex oxidation-reduction reactions are often time consuming and difficult to balance by inspection. One of the two common methods utilized to balance redox reactions and also elucidate the individual processes that occur at the two electrodes of a battery or electrolysis cell is called the method of half reactions.

In this technique, the redox reaction is divided into two half reactions, an oxidation and a reduction. A copper penny placed in a silver nitrate solution quickly picks up a coating of silver as the copper goes into solution according to the equation:

$$Cu_{(s)} + 2 Ag^+ = Cu^{2+} + 2 Ag_{(s)}$$

In this reaction copper is oxidized to copper ion and silver ion is reduced to elemental silver. The oxidation process, $Cu \rightarrow Cu^{2+}$, is first balanced according to mass and then electrons are added to balance charge:

$$Cu_{(s)} = Cu^{2+} + 2\ e^-$$

The same process is used for the reduction of silver ion to elemental silver, $Ag^+ \rightarrow Ag$.

$$Ag^+ + e^- = Ag_{(s)}$$

To obtain the net ionic equations, two half reactions are added together. As the silver reduction will have to occur two times for each copper oxidation, the silver half reaction is multiplied by two and added to the copper half reaction:

$$
\begin{array}{rcl}
2\ Ag^+ + 2\ e^- &=& 2\ Ag_{(s)} \\
Cu_{(s)} &=& Cu^{2+} + 2\ e^- \\
\hline
Cu_{(s)} + 2\ Ag^+ &=& Cu^{2+} + 2\ Ag_{(s)}
\end{array}
$$

Although this reaction could have easily been balanced by inspection, the half reaction method works for complex redox reactions when balancing by inspection can be extremely difficult.

Procedure

A. Metals as reducing agents.

Label four test tubes *1 - 4*.

1. Put a 1 cm² piece of copper metal into the first test tube.

2. Put a 2 cm long piece of magnesium ribbon into the second test tube.

3. Put a 1 cm² piece of zinc metal into a third test tube.

4. Put a small ball of steel wool into the fourth test tube.

5. Add a few milliliters of 0.1 M $CuSO_4$ to each tube. Note any reaction. Allow three minutes for a reaction to occur. If a reaction occurs, write a balanced equation. If not, write "NAR" for no apparent reaction.

6. Place the used metal squares in the strainer provided, rinse the tubes with deionized water, and test new metal squares in the same way as before with 3 M sulfuric acid. Hold a lighted splint over any test tube in which bubbling occurs.

7. Clean the tubes as before and test each new metal square with 0.1 M zinc sulfate solution.

B. Halogens and Fe^{3+} as oxidizing agents.

Clean out the test tubes and put 5 mL of 0.1 M sodium bromide in the first tube, 5 mL of 0.1 M sodium chloride in the second tube and 5 mL of 0.1 M sodium iodide in the third tube. Add 5 mL of 0.1 M $FeCl_3$ to each of the three tubes. Note any color change. A lightening in color resulting from a colored solution being diluted does not constitute a color change. Was the iron(III) able to oxidize any of the halides?

Clean out the test tubes and test 5 mL of each of the halide solutions with 5 mL of freshly prepared chlorine water. *Caution: This test must be performed in the hood.*

Prelaboratory Exercises - Experiment 22

For solutions to the starred problems, see *Appendix A*. For additional problem solving experience with oxidation numbers, half reactions and balancing oxidation-reduction reactions *Exercises 23, 24*.

Name_____Date_____Lab Section_____

1.* Elemental zinc reacts with aqueous nickel(II) chloride to give zinc chloride and nickel.

    a. Which is the more active metal (better reducing agent), zinc or nickel?

                                      _____

    b. Write oxidation and reduction half reactions and the net ionic reaction for the system.

2.* The following reaction does not occur spontaneously:

$$Pb^{2+} + 2 Cl^- = Pb_{(s)} + Cl_{2(aq)}$$

Which is the more active oxidizing agent, Pb^{2+} or Cl_2? _____

3. Elemental aluminum reacts with copper(II) sulfate to give aluminum sulfate and elemental copper. Write oxidation and reduction half reactions and the net ionic reaction for the system.

4. Elemental zinc reacts with hydrochloric acid in a single replacement reaction. Write oxidation and reduction half reactions and the net ionic reaction for the system.

5. Tin(II) nitrate reacts with silver nitrate to give tin(IV) nitrate and elemental silver. Write oxidation and reduction half reactions and the net ionic reaction for the system.

Results and Discussion - *Experiment 22*
OXIDATION-REDUCTION

Name_____Date_____Lab Section_____

A. Metals as reducing agents.

1. If an observable reaction occurs, complete and balance the equation. If there is not an observable reaction, write "NAR" (no apparent reaction). [Note: Assume that when elemental iron reacts, it ends up as iron(II).]

Balanced Equation

$CuSO_4(aq)$ + $Cu(s)$ = _____

$CuSO_4(aq)$ + $Mg(s)$ = _____

$CuSO_4(aq)$ + $Zn(s)$ = _____

$CuSO_4(aq)$ + $Fe(s)$ = _____

$H_2SO_4(aq)$ + $Cu(s)$ = _____

$H_2SO_4(aq)$ + $Mg(s)$ = _____

$H_2SO_4(aq)$ + $Zn(s)$ = _____

$H_2SO_4(aq)$ + $Fe(s)$ = _____

$ZnSO_4(aq)$ + $Cu(s)$ = _____

$ZnSO_4(aq)$ + $Mg(s)$ = _____

$ZnSO_4(aq)$ + $Zn(s)$ = _____

$ZnSO_4(aq)$ + $Fe(s)$ = _____

2. Arrange the four metals and hydrogen in order of their activities. Write the symbol of the best reducing agent first and the poorest one last.

1 (best) _____

2 _____

3 _____

4 _____

5 (poorest) _____

3. Write oxidation and reduction half reactions and the net ionic equations for:

a. copper(II) sulfate + zinc

   oxidation _____

   reduction _____

   net ionic _____

b. sulfuric acid + magnesium

   oxidation _____

   reduction _____

   net ionic _____

B. Halogens and Fe^{3+} as oxidizing agents.

1. If an observable reaction occurs, complete and balance the equation. If there is not an observable reaction, write "NAR" (no apparent reaction). [Note: Assume Fe^{3+} is reduced to Fe^{2+} when it reacts.]

_____Fe^{3+} + _____Br^- = _____

_____Fe^{3+} + _____Cl^- = _____

_____Fe^{3+} + _____I^- = _____

_____$Cl_2(aq)$ + _____Br^- = _____

_____$Cl_2(aq)$ + _____Cl^- = _____

_____$Cl_2(aq)$ + _____I^- = _____

2. Arrange Br_2, Cl_2, I_2 and Fe^{3+} in order of their reactivities. List the best oxidizing agent first and the poorest one last:

1 (best)   _____

2   _____

3   _____

4 (poorest)   _____

3. Write oxidation and reduction half reactions and the net ionic equations for:

a. iron(III)   +   iodide

   oxidation   _____

   reduction   _____

   net ionic   _____

b. chlorine   +   iodide

   oxidation   _____

   reduction   _____

   net ionic   _____

4. Suggest any ways you can think of to improve any part(s) of this experiment.

Experiment 23

ANALYSIS OF BLEACH AND COPPER(II) GLYCINATE

Learning Objectives

Upon completion of this experiment, students should have learned:
1. Iodometric techniques in quantitative analysis.
2. How to perform stoichiometric calculations involving solutions.

Text Topics

Stoichiometry of solution reactions, oxidation-reduction reactions (Malone, Chapters 11, 13)

Comments

The copper(II) glycinate to be used in this experiment should have been synthesized in *Experiment 20*. For problem solving experience involving oxidation-reduction reactions in solution and stoichiometry, see *Exercises 21, 24*.

Discussion

According to their labels, commercial household bleaches contain 5.25% sodium hypochlorite (NaClO). Today you will check the validity of that claim by titrating bleach with standardized sodium thiosulfate. The same titration technique will also be used to analyze the copper(II) glycinate you synthesized in *Experiment 20*.

A. Bleach analysis. The oxidation number of the chlorine in hypochlorite ion, the active ingredient in household bleach is +1. Because chlorine is more stable with an oxidation state of −1, the ClO⁻ is a strong oxidizing agent. It is the ability of ClO⁻ to remove electrons from colored compounds that results in its effectiveness as a bleaching agent. The hypochlorite ion reacts with iodide ion to yield elemental iodine and chloride ion:

$$H_2O_{(l)} + OCl^- + 2\,I^- = I_{2}(aq) + Cl^- + 2\,OH^-$$

In the presence of excess iodide ion , iodine further reacts to form the triiodide complex:

$$I_2 + I^- = I_3^-$$

The net reaction of hypochlorite with iodide is:

$$H_2O_{(l)} + OCl^- + 3 I^- = I_3^- + Cl^- + 2 OH^-$$

The triiodide is then reduced back to iodide by titration with a standardized thiosulfate solution:

$$I_3^- + 2 S_2O_3^{2-} = 3 I^- + S_4O_6^{2-}$$

The iodide ion thus does not undergo any net change in the process but assures that the correct stoichiometry occurs and provides a visually detectable endpoint. The overall reaction for calculation purposes is:

$$H_2O_{(l)} + 2 S_2O_3^{2-} + OCl^- = S_4O_6^{2-} + Cl^- + 2 OH^-$$

[Note: From the above equation we observe that the hypochlorite to thiosulfate mole ratio is 1:2]

Although the gradual disappearance of the triiodide color can be used to approximate the endpoint for the titration, a much sharper endpoint is achieved by the addition of a starch solution after the iodine color begins to fade. Iodine complexes with starch to form a dark purple color. When this color disappears, all of the triiodide has been reduced back to iodide.

Today, the reactions and techniques described above will be used to determine and compare the percent of NaClO in a brand name bleach and a generic brand of bleach. Using the molarity and volume of the sodium thiosulfate solution used to reach the endpoint, and the volume of bleach solution titrated, it is possible to calculate the percent of NaClO by mass in the bleach.

B. **Copper(II) glycinate monohydrate analysis.** A very similar procedure will be used to determine the percent by mass of copper in the product of the attempted synthesis of copper(II) glycinate monohydrate from **Experiment 20.** In addition, or alternatively, the mass percent of copper in an unknown can be determined using the technique below. The copper(II) compound is dissolved in dilute acid and reacted with excess potassium iodide. This results in the production of triiodide according to the following equation:

$$2 Cu^{2+} + 5 I^- = 2 CuI_{(s)} + I_3^-$$

The triiodide produced in this reaction is titrated with standardized thiosulfate as in the bleach titration:

$$I_3^- + 2 S_2O_3^{2-} = 3 I^- + S_4O_6^{2-}$$

and the overall reactions is:

$$2 Cu^{2+} + 2 I^- + 2 S_2O_3^{2-} = 2 CuI_{(s)} + S_4O_6^{2-}$$

[Note: From the above equation, we observe that the copper(II) to thiosulfate mole ratio is 1:1]

Procedure

   A. Bleach analysis. This analysis should be performed on two different brands of bleach; preferably one name brand and one generic. Pipet 10.00 mL of the first bleach into a 100 mL volumetric flask. Dilute to the 100 mL mark with deionized water and mix well. Rinse the pipet out with a little of the diluted bleach solution twice and pipet a 10.00 mL aliquot of it into a 250 mL Erlenmeyer flask. Add 1.0 g of potassium iodide and swirl the resulting mixture. Now add 5.0 mL of 6 M HCl to the mixture. *[Caution: Make sure you add potassium iodide to the bleach solution <u>before</u> you add the hydrochloric acid. Addition of the acid first results in liberation of poisonous chlorine gas.]*

   Titrate with the standardized sodium thiosulfate (~0.11 M) solution until the amber iodine color begins to fade. At this point, add 2 mL of 0.4% starch solution and resume titrating until the dark color of the starch-iodine complex just disappears. Be sure to use ½ drop quantities as you near the endpoint by swiftly turning the stopcock 180°. An abrupt color change from dark blue to colorless marks the endpoint. Repeat the titration on a second 10.00 mL aliquot of the diluted bleach solution. Repeat the analysis with a second brand of bleach.

   B. Copper(II) glycinate monohydrate analysis. Accurately weigh out about 0.3 g to at least the nearest 0.01 g of the copper(II) glycinate monohydrate (from Experiment 20) or a copper(II) unknown into a 250 mL Erlenmeyer flask and dissolve it in 35 mL of 0.05 M sulfuric acid. Add 1.0 g of potassium iodide and swirl the mixture. Titrate the mixture with standardized sodium thiosulfate until the triiodide color begins to fade. At this point, add 2 mL of 0.4% starch solution. Continue to titrate until the gray starch-iodine color disappears. An abrupt transition from a gray-yellow color to a light pink color will mark the endpoint. Repeat the titration on a second 0.3 g sample of the complex or unknown. Calculate the percent by mass of copper in your sample. If you analyzed the synthesis product, copper(II) glycinate monohydrate, calculate the mass percent of copper from the formula and compare the theoretical and experimental values. Was your synthesis successful?

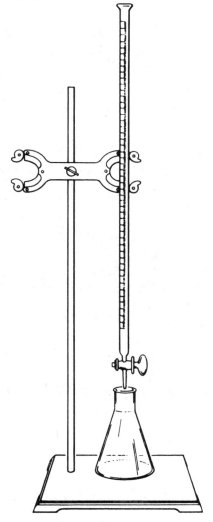

Figure 23-1

Chemical Capsule

Part of this experiment involves the determination of the amount of sodium hypochlorite (according to the label it is 5.25% by mass) in bleach. Bleach is just one of the many chemicals found in most households. Many of these chemicals including bleach are potenially very dangerous as they are corrosive and very toxic.

Many household cleaning agents contain ammonia. One should be sure that bleach and ammonia products are never mixed as they react according to equations including the following:

$$NH_3(aq) + NaOCl(aq) \rightarrow NH_2Cl(aq) + NaOH(aq)$$

$$2\,NH_3(aq) + NaOCl(aq) \rightarrow N_2H_4(aq) + NaCl(aq) + H_2O(l)$$

Both the hydrazine and chloroamine produced are very toxic.

Another potentially hazardous situation can occur with the chemicals used to prevent bacterial growth in pools and to adjust the pH. One method of chlorinating pools is to use a sodium hypochlorite solution that is usually supplied by pool stores at twice the concentration of household bleach. If the pH of the pool is too high, muriatic acid (commercial name for hydrochloric acid) is usually added to the pool. If NaOCl solution and HCl are poured into the same location of the pool in a short amount of time, toxic chlorine gas will be produced.

$$NaOCl(s) + 2\,HCl(aq) \rightarrow Cl_2(g) + NaCl(aq) + H_2O(s)$$

The two solutions should be added to different ends of the pool preferably at different times.

Prelaboratory Exercises - *Experiment 23*

For solutions to the starred problems, see *Appendix A*. For problem solving experience involving oxidation-reduction reactions in solution and stoichiometry, see *Exercises 21, 24.*

Name_____Date_____Lab Section_____

1.* A 1.00 mL sample of pool "chlorine" (NaClO, density = 1.18 g/mL) required 21.12 mL of 0.150 M $Na_2S_2O_3$ to titrate it to the endpoint using the iodometric technique. What was the molarity and percent of sodium hypochlorite in the pool chlorine?

_____

_____

2.* A 0.250 g sample believed to be copper(II) acetate monohydrate required 10.36 mL of 0.1200 M $Na_2S_2O_3$ to reach the endpoint in an iodometric titration. Calculate the experimental percent by mass of copper in the sample and the theoretical mass percent of copper in copper(II) acetate monohydrate.

_____

_____

3. Iodine interacts with starch to form a purple colored complex. Explain why the end point is reached in these titrations when the purple color disappears.

4. A 5.00 mL sample of aqueous sodium hypochlorite required 18.30 mL of 0.1250 M $Na_2S_2O_3$ to titrate it to the endpoint using the iodometric technique. What was the molarity of sodium hypochlorite in the solution?

    _____

5. A 0.205 g sample believed to be copper(II) chloride dihydrate required 11.46 mL of 0.1330 M $Na_2S_2O_3$ to reach the endpoint in an iodometric titration. Calculate the experimental percent by mass of copper in the sample and the theoretical mass percent of copper in copper(II) chloride dihydrate.

    _____

    _____

Results and Discussion - *Experiment 23*
ANALYSIS OF BLEACH AND COPPER(II) GLYCINATE

Name_____Date_____Lab Section_____

A. Bleach Analysis

1. Molarity of $Na_2S_2O_3$ solution (from bottle)   _____

		Bleach 1		Bleach 2	
2. Brand		_____		_____	
3. Final buret reading		_____	_____	_____	_____
4. Initial buret reading		_____	_____	_____	_____
5. Volume $Na_2S_2O_3$ solution		_____	_____	_____	_____
6. Moles of $Na_2S_2O_3$		_____	_____	_____	_____
7. Moles NaClO in 10 mL aliquot[1]		_____	_____	_____	_____
8. Molarity of NaClO in 10 mL aliquot[2]		_____	_____	_____	_____
9. Molarity of NaClO in bleach		_____	_____	_____	_____
10. Average molarity of NaClO in bleach		_____		_____	
11. Grams/L of NaClO in bleach		_____		_____	
12. Mass percent of NaClO in bleach[3]		_____		_____	
13. Mass percent of NaClO from manufacturer's label		_____		_____	
14. Percent deviation between exptl. and label values		_____		_____	

---

[1]Be sure to take into account the correct mole ratio!

[2]This value equals the molarity of NaClO in the 100 mL volumetric flask.

[3]Assume the density of bleach is 1.08 g/mL

267

15. Show the series of unit conversions you could use to calculate the molarity of the NaClO in the 100 mL flask from the volume and molarity of the $Na_2S_2O_3$ solution, the molar ratio and the volume of the NaClO for titration 1 of the first bleach.

16. Show the series of unit conversions you could use to calculate the mass percent of NaClO in bleach 1 from the average molarity, the formula mass of NaClO and the density of bleach.

B. Copper(II) glycinate monohydrate analysis

1. Molarity of $Na_2S_2O_3$ solution (from bottle) _____

	Sample 1	*Sample 2*
2. Mass of copper glycinate	_____	_____
3. Final buret reading	_____	_____
4. Initial buret reading	_____	_____
5. Volume $Na_2S_2O_3$ solution	_____	_____
6. Moles of $Na_2S_2O_3$	_____	_____
7. Moles of Cu^{2+}	_____	_____
8. Grams of Cu^{2+}	_____	_____
9. Mass percent of copper in sample	_____	_____

10. Average mass percent of copper in sample _____

11. Theoretical mass percent of copper in sample _____
    (formula mass of copper(II) glycinate monohydrate = 229.66 g/mol)

12. Percent error _____

13. Show the series of unit conversions you could use to calculate the mass percent of copper in copper(II) glycinate monohydrate for sample 1 from the volume and molarity of the $Na_2S_2O_3$ solution, the mass of the copper(II) glycinate monohydrate, the molar ratio and the atomic mass of copper.

Experiment 24

THE RATES OF CHEMICAL REACTIONS

Learning Objectives

Upon completion of this experiment, students should have learned:
1. The effects of concentration, temperature and catalysts on reactions rates.
2. The general form of a rate expression.
3. Aspects of reaction mechanisms.

Text Topics

Reaction rates and mechanisms (Malone, Chapter 14)

Comments

While this experiment does demonstrate some of the general principles of the field of kinetics, the techniques employed in this experiment lead to approximate results only and should not be considered rigorously correct.

Discussion

The rates of many processes are of interest to us. The speed of your car in miles per hour, the number of kilowatts of electricity used in your home per month or the cost per minute for a phone call. Notice the rate of each of these processes is expressed in units of some quantity (distance, energy or dollars) divided by a time period. The rates of chemical reactions expressed in the change in the amount of a chemical reactant (often moles/liter) divided by time (usually seconds) are of considerable significance in practical and theoretical aspects of chemistry. From a practical viewpoint, it is desirable to adjust conditions for a reaction such that the reaction occurs in a reasonable amount of time (minutes, hours or even a few days). It should not be so slow that one has to wait years or so fast that an explosion results. From a theoretical perspective, a study of the variation of reaction rates with respect to concentration and temperature helps elucidate mechanisms of reactions.

Temperature effects. The rate of virtually every reaction increases as the temperature increases (as a rough rule of thumb, a 10°C increase doubles the reaction rate). Consider the potentially explosive reaction between hydrogen and oxygen to yield water. Although the reaction is highly exothermic, a mixture of hydrogen and oxygen will sit in a bottle indefinitely without noticeable formation of water. Before the oxygens can begin to form bonds with hydrogens, the hydrogen-hydrogen and oxygen-oxygen bonds must begin to break. The cleavage of bonds requires energy. Apparently this process requires more energy than most of the molecules have at ambient temperature. As the temperature of the system is increased, the fraction of molecules with sufficient energy to undergo bond cleavage upon a collision increases dramatically and the rate of reaction increases.

Compare the situation to a bowl containing several marbles. If the bowl is lifted off the ground, the marbles will have potential energy because of their position relative to the ground. But because of the lip of the bowl, the marbles are hindered from returning to the lower energy state. If energy is transferred to the system by shaking the bowl, eventually, if the shaking is hard enough, some of the marbles will attain enough energy to make it over the lip and fall to the ground. In this case, the energy of the marble is transferred to the ground. In the case of the molecules, the energy that is released when product bond formation occurs can often be transferred to unreacted molecules to help them break bonds or overcome the energy of activation barrier. Some reactants, such as the hydrogen oxygen mixture, will react on their own after enough molecules have overcome the energy barrier. Other reactions need continual assistance. This is almost always true of endothermic reactions since not enough usable energy is released to help the molecules over the energy barrier.

<u>Concentration effects.</u> The dependence of the rate of reaction on the concentration of reactants is more intuitively understandable. An increase in the concentrations of the reactants will often increase the reaction rate. Consider a reaction between hydrogen and iodine to produce hydrogen iodide ($H_2 + I_2 = HI$). For the reaction to proceed, a collision between H_2 and I_2 must occur. One would expect that a doubling of either the H_2 concentration or the I_2 concentration would double the number of collisions and the rate. A doubling of both H_2 and I_2 should quadruple the number of collisions and the rate. Remember that the rate is expressed as a change in the amount of a reactant, $\Delta[H_2]$ per unit time, t, or rate = $\Delta[H_2]/\Delta t$. In this reaction, the rate should be proportional to the concentrations of H_2 and I_2 or:

$$\Delta[H_2]/\Delta t = -k[H_2][I_2]$$

where k is the proportionality constant that is called the rate constant. The above rate expression is consistent with the conclusions above that a doubling of the concentration of either hydrogen or iodine should double the rate. It should be noted that the rate constant, k is temperature dependent and increases markedly with temperature due to an increase in the number and more importantly, the effectiveness of the collisions. The rate expression then for **any step** of a reaction will be proportional to the products of the concentrations of the reacting species each raised to a power given by the coefficient of the species in the reaction step. Some examples are:

A	$\rightarrow$ P	$\Delta[A]/\Delta t = -k[A]$
2 A	$\rightarrow$ P	$\Delta[A]/\Delta t = -2k[A]^2$
A + B	$\rightarrow$ P	$\Delta[A]/\Delta t = -k[A][B]$

If the reaction is a multistep reaction, the determination of the rate expression involves a somewhat more complex analysis that is beyond the scope of this discussion. In fact, the system you will study today is a multistep process but the possible rate expressions will be provided to you.

Catalysts. Some additives to a reaction mixture markedly increase the reaction rate by providing alternate or lower energy pathways to products. Compounds that increase reaction rates without undergoing any *net* chemical change themselves are called catalysts.

Multistep reactions. The stoichiometry of the reaction you will explore today is :

$$2 \text{ I}^- + \text{S}_2\text{O}_8{}^{2-} = \text{I}_2\text{(aq)} + 2 \text{ SO}_4{}^{2-}$$

This reaction apparently requires the simultaneous collision of three ions. The probability of such an occurrence is very small. Consider the probability of John finding Mary and Jane at the same time. The chance of such a coincidence is small. However, John could encounter one of them and then the two of them could go looking for the third person. Reactions that involve more than two molecules or ions also usually proceed in steps.

For the reaction between two iodides and a persulfate, a possible sequence would be the following:

$$\text{I}^- + \text{S}_2\text{O}_8{}^{2-} = \text{SO}_4{}^{2-} + \text{SO}_4\text{I}^-$$

mechanism 1

$$\text{SO}_4\text{I}^- + \text{I}^- = \text{I}_2\text{(aq)} + \text{SO}_4{}^{2-}$$

One of these reactions would probably be slower than the other. The slower reaction in a multistep reaction is called the rate determining step. This means that the overall rate for the process is determined primarily by the rate of the slow step. By assuming which step is the rate determining step and making an approximation, it is possible to derive a rate expression for the reaction. For *mechanism 1*, if the first step is rate determining, then the rate expression is just the rate expression for the first step.

$$\text{rate} = k[\text{I}^-][\text{S}_2\text{O}_8{}^{2-}] \qquad \text{rate expression for mechanism 1-a}$$

If the second step is slower than the first, the rate expression comes out in the same form as though the reaction occurs in one step.

$$\text{rate} = k[\text{I}^-]^2[\text{S}_2\text{O}_8{}^{2-}] \qquad \text{rate expression for mechanism 1-b and for improbable one step reaction}$$

A second possible sequence is:

$$2 \text{ I}^- = \text{I}_2{}^{2-}$$

mechanism 2

$$\text{I}_2{}^{2-} + \text{S}_2\text{O}_8{}^{2-} = \text{I}_2\text{(aq)} + 2 \text{ SO}_4{}^{2-}$$

If the first step is the rate determining step:

$$\text{rate} = 2k[\text{I}^-]^2 \qquad \text{rate expression for mechanism 2-a}$$

If the second step is the rate determining step, the rate expression is the same as if the reaction occurred in one step.

$$\text{rate} = k[\text{I}^-]^2[\text{S}_2\text{O}_8{}^{2-}] \qquad \text{rate expression for mechanism 2-b}$$

Your experimental results should enable you to distinguish among the three rate expressions and eliminate some of the possible mechanisms. Rate studies cannot prove that a proposed mechanism is operative but only disprove possible mechanisms.

The solutions for each of the kinetic runs are given in the table below. The sodium thiosulfate ($Na_2S_2O_3$) is used as part of an indicator system (a purple color will form) to enable you to determine the time required for the reaction to occur for each kinetic run. As the amount of product formed in each case (SO_4^{2-}) will be the same, **the relative rates of reaction will be inversely proportional to the reaction time**. For example, a reaction that takes 25 seconds has twice the rate of one that takes 50 seconds. Because of the proportionality, for the analysis today, it will be possible to substitute the inverse of the time (1/t) for the actual rate of the reaction.

To study the effect of reactant (I^- and $S_2O_8^{2-}$) concentrations on reaction rates, the reactant concentrations will be decreased. The simplest method for achieving this goal is to use a smaller amount of one of the reactants while keeping the total volume constant by replacement with water containing a substance that will not affect the reaction but that will maintain the same net ionic strength. Solutions 3 and 4 are used for this purpose.

Solution number	Contents
1	0.20 M KI
2	0.0050 M $Na_2S_2O_3$ in a 0.4% starch solution
3	0.2 M KCl
4	0.1 M K_2SO_4
5	0.1 M $CuSO_4$
6	0.10 M $K_2S_2O_8$

Procedure

A. Concentration effects on reaction rates. For each run, add with a pipet or 25 mL graduated cylinder or buret, the required amounts of solutions 1 - 5 to a 125 mL or 250 mL Erlenmeyer flask. Measure the required amount of solution 6 and add it to another flask. Quickly add the contents of the flask containing the 0.10 M $K_2S_2O_8$ (solution #6) to the first flask, swirl, and start timing. Record the time when the purple appears. After the purple appears, measure and record the temperature of each run. Note that run #1 is performed three times to check for the precision of the experiment. The other runs vary the concentration of one of the reactants or include a catalyst.

Milliliters of each solution needed for each kinetic run

Solution Number				Run Number						
	1a	*1b*	*1c*	*2*	*3*	*4*	*5*	*6*	*7*	*8*
1	20	20	20	15	10	5	20	20	20	20
2	10	10	10	10	10	10	10	10	10	10
3	-	-	-	5	10	15	-	-	-	-
4	-	-	-	-	-	-	5	10	15	-
5	-	-	-	-	-	-	-	-	-	1 drop
6	20	20	20	20	20	20	15	10	5	20

B. Temperature effects on reaction rates. To study the effect of temperature on the reaction rate, run *1* will be repeated at several temperatures. The flask containing solutions *1* and *2* will be heated or cooled and then solution *6* (at room temperature) will be added. It will be assumed that quick mixing will result in the same temperature for the whole solution. The temperature will be measured immediately after the purple color appears.

For each run, add 20 mL of solution *1* and 10 mL of solution *2* to the first flask. Add 20 mL of solution *6* to the second flask. Change the temperature of the first flask to about the value indicated below. To raise the temperature of a flask, suspend the flask in a beaker of water and heat the beaker over a wire gauze with a Bunsen burner. To lower the temperature of a flask, swirl the solution in an ice bath until the desired temperature is achieved. Quickly add solution *6*, commence timing and stir. Record the time elapsed for the purple to appear and measure the temperature of the solution. Repeat the procedure for the remaining runs.

run number	approximate temperature for flask 1 (°C)
1a, 1b, 1c	room temperature (already performed above)
2'	40
3'	5

Chemical Capsule

Looking around today at the human made objects all around us, we notice that a significant number are composed of various kinds of polymers. From our polyester clothes to styrofoam coffee cups, we are surrounded by a world of polymeric compounds. Some people argue that styrofoam and other polystyrene products are damaging the environment because they are not biodegradable. The options available are not necessarily better. Production of paper uses trees, considerable energy and paper bleaching can result in the production of toxic chemicals. Also it has been found that paper does not biodegrade in the kinds of landfills we currently use. Reusables such as plastic or porcelain dishes offer many advantages over their styrofoam and paper counterparts but even reusables cause environmental problems. It takes water and potentially harmful detergents to clean the dishes. The message is that issues like these are much more complicated than it might appear and the best solution may vary from one situation to the next. Most important, consider all the issues.

Another common polymer, polyvinyl chloride (PVC) is used for many applications including water pipes and seat covers. The compound used to make PVC, vinyl chloride (C_2H_3Cl), is highly carcinogenic. Before this was recognized many people involved in the industrial preparation of PVC unfortunately acquired liver cancer.

Despite occasional very unfortunate mistakes like the one with vinyl chloride, the chemical industry is one of the safest of all industries for its workers in terms of days lost from the job.

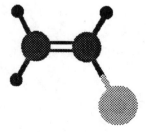

vinyl chloride

Prelaboratory Exercises - *Experiment 24*

For solutions to the starred problems, see *Appendix A*.

Name_____Date_____Lab Section_____

1.* Consider the hypothetical reaction: A + B + C = P

The time required to produce the same small amount of P for several concentrations of A, B and C are given in the table below:

run #	1	2	3	4	5	6	7
A (moles/L)	0.10	0.050	0.025	0.10	0.10	0.10	0.10
B (moles/L)	0.10	0.10	0.10	0.050	0.025	0.10	0.10
C (moles/L)	0.10	0.10	0.10	0.10	0.10	0.050	0.025
t (seconds)	40	78	163	38	42	82	158
1/t (1/seconds)	_____	_____	_____	_____	_____	_____	_____

Determine the relative rate constants (1/t) for each solution and the rate law for the reaction. Suggest a stepwise mechanism for the reaction.

2. For a reaction with the stoichiometry, A + 3 B = 2 P, write an expression that relates the rate of disappearance of A to the rate of disappearance of B and another expression that relates the rate of disappearance of A to the rate of appearance of P.

3. The *Procedure section* states "As the amount of product formed in each case (SO_4^{2-}) will be the same, **the relative rates of reaction will be inversely proportional to the reaction time**." Explain this statement using math if appropriate.

4.* What is the function of solution 2 in this experiment?

5.* What are the functions of solutions 3 and 4 in this experiment?

6. What is the purpose of using solution 5 in this experiment?

Results and Discussion - *Experiment 24*
THE RATES OF CHEMICAL REACTIONS

Name_____Date_____Lab Section_____

A. Concentration effects on reaction rates.

1. Concentrations and results.

run #	[I⁻] (mol/L)	[S₂O₈²⁻] (mol/L)	time (sec.)	1/time (sec.⁻¹)	temp. (°C)
1a	_____	_____	_____	_____	_____
1b	_____	_____	_____	_____	_____
1c	_____	_____	_____	_____	_____
2	_____	_____	_____	_____	_____
3	_____	_____	_____	_____	_____
4	_____	_____	_____	_____	_____
5	_____	_____	_____	_____	_____
6	_____	_____	_____	_____	_____
7	_____	_____	_____	_____	_____
8	_____	_____	_____	_____	_____

2. Considering the results of *1a, 1b, and 1c,* how reproducible are the results of this experiment?

3. Is the rate of the reaction proportional to the iodide concentration?
   Explain your answer.   _____

4. Is the rate of the reaction proportional to the persulfate concentration? Explain your answer. _____

5. Write down the probable steps and the rate expression for the reaction.

6. Did the copper(II) sulfate catalyze the reaction? Explain your answer. _____

B. Temperature effects on reaction rates.

1. Results

run #	time (sec.)	1/time (sec.$^{-1}$)	temp. (°C)
1 (average of *1a, 1b, 1c*)	_____	_____	_____
2'	_____	_____	_____
3'	_____	_____	_____

2. For this particular reaction, how good is the rule of thumb that a "10°C increase should double the reaction rate? Explain your answer.

3. Suggest any ways you can think of to improve any part(s) of this experiment.

Experiment 25

EQUILIBRIUM STUDIES

Learning Objectives

Upon completion of this experiment, students should have learned:
1. Le Chatelier's Principle.
2. The nature of a complex.
3. Concepts of acid-base indicators.
4. How to write expressions for the equilibrium constant.

Text Topics

Le Chatelier's Principle, equilibrium constant expressions (Malone, Chapter 14)

Comments

For additional problem solving experience with equilibrium concepts, see *Exercise 25.*

Discussion

Many chemical reactions are reversible. In other words, if two chemical species in solution are mixed and form new species, there is some tendency for the new species to react reforming the original chemical species. The rate of formation of the new species will, for a time, be faster than the reverse reaction. However, when the visible reaction ceases, the rate of formation of the new species becomes equal to the rate of the reverse reaction and a state of equilibrium is said to be reached. Equilibrium equations are written with two arrows pointed in opposite directions between reactants and products, indicating that both processes are taking place simultaneously.

Reactants $\rightleftarrows$ Products

For example, the equilibrium for saturated silver chloride (undissolved AgCl remains in the bottom of the test tube) is represented by the expression:

$$AgCl_{(s)} \rightleftharpoons Ag^+ + Cl^-$$

Solid silver chloride is breaking into silver and chloride ions at the same rate that silver ions and chloride ions are coming together to reform solid silver chloride.

Acetic acid slightly dissociates according to:

$$HC_2H_3O_{2(aq)} \rightleftharpoons H^+ + C_2H_3O_2^-$$

and equilibrium is achieved when the rate of the forward reaction equals the rate of the reverse reaction. For systems for which a state of equilibrium exists, it can be shown that the product of the concentrations of the products divided by the product of the concentrations of the reactants is equal to a constant. For example, for a system with two reactants, A and B, and two products, C and D;

$$A + B \rightleftharpoons C + D \qquad K_{eq} = \frac{[C][D]}{[A][B]}$$

The mathematical relationship for K_{eq} demonstrates Le Chatelier's Principle that a stress placed on a system will cause the system to shift in a direction to reachieve equilibrium. If an aqueous acetic acid solution

$$\text{for which} \quad K_a = \frac{[H^+][C_2H_3O_2^-]}{[HC_2H_3O_2]}$$

is disturbed by the addition of HCl, the system will shift to the left as the hydrogen ion concentration has increased. Looking at the equilibrium expression, addition of H^+ makes the numerator large relative to the denominator. Consequently, some H^+ must react with $C_2H_3O_2^-$ to decrease the numerator and increase the denominator until the hydrogen ion concentration multiplied by acetate concentration divided by the acetic acid concentration once again equals K_a. Addition of NaOH to the system would decrease the hydrogen ion concentration and the system would shift to the right to reestablish equilibrium.

Solubility Products:

The equilibrium constant for compounds that have very low solubilities in water is called a solubility product. For example, only 2×10^{-4} grams of silver chloride dissolves in 100 mL of water to give silver and chloride ions:

$$AgCl_{(s)} \rightleftharpoons Ag^+ + Cl^- \qquad K_{eq} = \frac{[Ag^+][Cl^-]}{[AgCl]}$$

$$\text{or} \quad K_{eq}[AgCl] = [Ag^+][Cl^-]$$

Since the concentration of solid AgCl is a constant, the product, $K_{eq}[AgCl]$, can be defined by a new constant, the solubility product, K_{sp}.

$$K_{sp} = [Ag^+][Cl^-]$$

Calcium hydroxide also has a low solubility in water (0.1 g/100 mL). The dissociation of calcium hydroxide into ions can be represented by:

$$Ca(OH)_{2(s)} \rightleftharpoons Ca^{2+} + 2 OH^-$$

For the system above which will be studied in today's experiment:

$$K_{sp} = [Ca^{2+}][OH^-]^2$$

This means that the concentration of one ion is dependent on the concentration of the other ion.

Complex ions:

Copper(II) ions react with ammonia to form a complex ion according to the equation:

$$Cu^{2+} + 4 NH_{3(aq)} \rightleftharpoons Cu(NH_3)_4^{2+}$$

A decrease in the concentration of either reactant should cause the complex to dissociate into reactants in order reestablish equilibrium.

Indicator equilibria:

Indicators are very useful additives in acid-base titrations for the visualization of the endpoint. Indicators are themselves often acids of the general formula HIn. In a solution an indicator is in equilibrium with its conjugate base according to:

$$HIn_{(aq)} \rightleftharpoons H^+ + In^-$$

The position of equilibrium varies with the acidity of the solution (or the pH) and is different at any pH value for each indicator. For compounds that are used as indicators, the color of HIn and In$^-$ are different. As the system shifts from a preponderance of HIn to a preponderance of In$^-$ over a narrow pH range (usually about 1.5 units), it is possible to detect pH changes by observing color changes of an appropriate indicator. For example, when titrating a strong acid with a strong base, an indicator which changes color around a pH of 7 is used to detect the neutralization point. In this section, you will study the shift in the position of equilibrium for the indicator thymol blue as acid or base is added to the system.

Procedure

A. Solubility of calcium hydroxide.

1. To 5 mL of a saturated solution of calcium hydroxide, add 5 mL of 6 M sodium hydroxide. Report and explain your observations.

2. To the above mixture, add 5 mL of 1.0 calcium chloride. Report and explain your observations.

3. Now add 8 mL of 6 M HCl to the above mixture. Stopper the tube and mix the contents. Report and explain your observations.

B. Equilibrium studies involving the copper(II) ammonia complex.

1. Mix 3 mL of a 0.1 M $CuSO_4$ solution and 3 mL of 6 M NH_3 together in a test tube. Report and explain your observations.

2. To the above mixture add 3 mL of 6 M HCl to the system. Report and explain your observations.

C. Equilibrium studies involving indicators.

1. Add 10 drops of thymol blue indicator to 10 mL of deionized water in a 20 × 150 mm test tube. Mix and record the color of the system.

2. Add 0.010 M NaOH to the above mixture drop by drop while mixing and record your observations. Then add 0.012 M HCl drop by drop and record your observations.

3. Thymol blue exhibits the rather unusual property of undergoing a second color change at a very low pH. Add 1.0 M HCl drop by drop, with mixing, to the thymol blue solution. Report your observations.

Prelaboratory Exercises - *Experiment 25*

For solutions to the starred problems, see *Appendix A*. For additional problem solving experience with equilibrium concepts, see *Exercise 25*.

Name_____Date_____Lab Section_____

1. Write equilibrium constant expressions for the following reactions:

    a.* $PCl_5(g) = PCl_3(g) + Cl_2(g)$ b. $2 NO(g) = N_2(g) + O_2(g)$

2.* Write the expression for the solubility product for the dissolving of $CaSO_4$ in water:

    a. $CaSO_4(s) = Ca^{2+} + SO_4^{2-}$

    What would you expect to observe in a saturated $CaSO_4$ solution if:

    b. $Ca(NO_3)_2$ is added?

    c. H_2SO_4 is added?

    d. NaCl is added?

3. For the reaction of ammonia with water:

$$NH_3(aq) + H_2O(l) = NH_4^+ + OH^-$$

in which direction (right or left) would the system shift to achieve equilibrium if:

a. NaOH is added to the solution?   _____

b. HCl is added to the solution?   _____

c. NH_3 evaporates out of the solution?   _____

Results and Discussion - *Experiment 25*
EQUILIBRIUM STUDIES

Name_____Date_____Lab Section_____

A. Solubility of calcium hydroxide.

1. Write the expression for the solubility product for the dissolving of calcium hydroxide in water: $Ca(OH)_2(s) = Ca^{2+} + 2\ OH^-$

$$K_{sp} = \underline{\hspace{3cm}}$$

2. Observations when 6 M NaOH was added to saturated calcium hydroxide:

Explanation:

3. Observations when 1.0 M $CaCl_2$ was added to solution from *A-2*:

Explanation:

4. Observations when 6 M HCl was added to the mixture from *A-3*?

Explanation:

B. Equilibrium studies involving the copper(II) ammonia complex.

1. Write the equilibrium expression for the reaction of copper(II) and ammonia:

$Cu^{2+} + 4\ NH_3(aq) \rightleftharpoons Cu(NH_3)_4^{2+}$

$$K_{sp} = \underline{\hspace{3cm}}$$

2. Observations when 0.1 M $CuSO_4$ is mixed with 6 M NH_3:

Explanation:

C. Equilibrium studies involving indicators.

1. Write the equilibrium expression for the dissociation of thymol blue:
   $HIn_{(aq)} = H^+ + In^-$

   $K = $ _____

2. Color of thymol blue in deionized water: _____

3. Observations on addition of 0.010 M NaOH and 0.012 M HCl:

# of drops of NaOH	color	# of drops of HCl	color
1	_____	1	_____
2	_____	2	_____
3	_____	3	_____
4	_____	4	_____
5	_____	5	_____
6	_____	6	_____
7	_____	7	_____
8	_____	8	_____
9	_____	9	_____
10	_____	10	_____

3. What was the predominant species in acid solution, HIn or In^-? _____

4. What was the predominant species present in base solution, HIn or In^-? _____

5. Did you observe an intermediate color for the indicator between the two extremes? If so, how do you explain it? _____

6. Observations when 1.0 M HCl was added to the system:

Experiment 26

MOLECULAR MODELS OF ORGANIC COMPOUNDS

Learning Objectives

Upon completion of this experiment, students should have learned:
1. Geometric aspects of organic molecules.
2. Structural, geometric and stereoisomerism.
3. The importance of chirality in nature.
4. To utilize molecular models.

Text Topics

Structure of organic molecules, bond angles, isomerism (Malone, Chapters 16)

Comments

Students should review *Experiment 7* before starting this experiment. For problem solving experience with organic nomenclature, functional groups and bond angles, see *Exercise 27*. Drawing the Lewis structures before the beginning of the laboratory period will save considerable laboratory time and increase the educational value of the experiment.

Discussion

The 3-dimensional nature of most molecules plays a significant role in determining physical and chemical properties of the substance. For example, a substrate must exactly fit into a portion of an enzyme if the enzyme is to perform its catalytic function just as a key must have just the right shape to fit a lock. You have probably discussed the VSEPR (valence shell electron pair repulsion) model and observed that the number of "groups" (neighboring atoms plus non-bonded electron pairs) around a central atom leads to an accurate prediction of bond angles (2 groups - 180°, 3 groups - 120°, 4 groups - 109°). The molecular models you will build in today's exercise go a step further and enable you to visualize the spatial arrangement of the entire molecule. In some cases, this should help you understand an extremely important concept in organic chemistry called isomerism. Before you start this exercise, look up the definition of isomerism and familiarize yourself with the concept.

The importance of the sequence and spatial arrangement of the atoms is established with the concept of isomerism. Consider the formula $C_2H_2Cl_2$. It is possible to draw three acceptable Lewis structures for this formula that correspond to the compounds 1,1-dichloroethene, *cis*-1,2-dichloroethene and *trans*-1,2-dichloroethene.

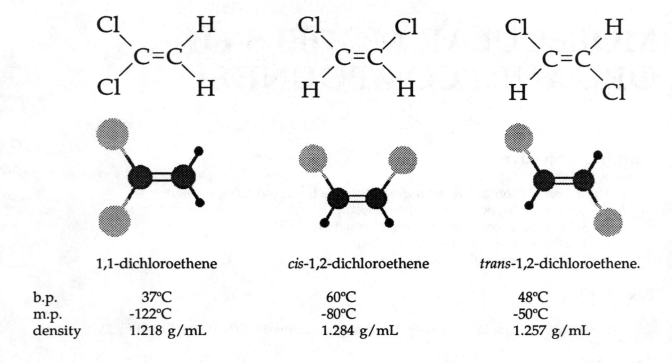

	1,1-dichloroethene	*cis*-1,2-dichloroethene	*trans*-1,2-dichloroethene.
b.p.	37°C	60°C	48°C
m.p.	-122°C	-80°C	-50°C
density	1.218 g/mL	1.284 g/mL	1.257 g/mL

The three possible **isomers** exist and as can be observed from the data above, have different properties. This demonstrates the fact that the spatial arrangement of the atoms has significant effects on the physical and chemical properties of the molecule. The first isomer (1,1-dichloroethene) has a different sequence of bonding than the other two and is a structural isomer of the other two. The second and third isomers have the same sequence of bonding but are spatially different. Because they differ in geometry, they are called geometric isomers. Notice that because of the presence of the double bond, rotation from the *cis* to the *trans* is not possible unless substantial energy input is provided. The absorption of light by rhodopsin in the eye causes a *cis - trans* isomerization that leads to the nerve impulse that is sent to the brain and results in vision.

Procedure

Review the concepts of VSEPR and the *Discussion* section in *Experiment 7*.

A. Construct a model of a methane molecule (CH_4). Observe and record the bond angles.

B. Make a model of ethane (C_2H_6). Observe the free rotation about the carbon - carbon single bond. The different rotational positions (eclipsed, staggered) are called conformers. Record the bond angles.

C. Make two models of propane (C_3H_8). Again notice the rotational possibilities and record the bond angles. Compare the structure to the Lewis structure below. Remember that the Lewis structures of 3-dimensional molecules will not correctly show bond angles and the 90° bond angles typically included in Lewis structures are seldom correct.

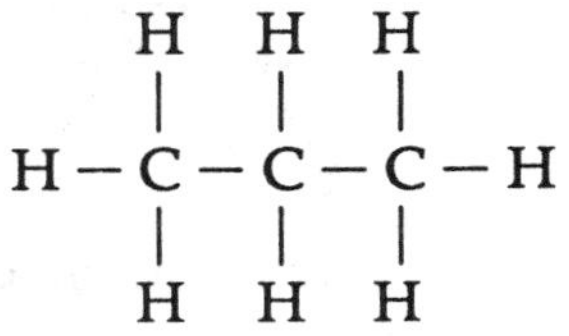

```
          H   H   H
          |   |   |
      H — C — C — C — H
          |   |   |
          H   H   H
```

Although the Lewis structure of propane is technically the figure on the left, it is sometimes more convenient to use partially condensed formulas, $CH_3 — CH_2 — CH_3$ or even fully condensed formulas, $CH_3CH_2CH_3$.

D. Remove one of the hydrogens from the first carbon on one of the propane models and replace it with a methyl group ($— CH_3$). This is a model of butane. With the other propane model, replace a hydrogen on the second carbon with a methyl group. This compound is 2-methylpropane (its common name is isobutane). Notice both models have the same formula C_4H_{10}, but different structures and therefore different chemical and physical properties. Look up the boiling and melting points of the two isomers in the *Handbook of Chemistry and Physics*. The fully condensed formulas for the two isomers are $CH_3CH_2CH_2CH_3$ (*n*-butane) and $CH_3CH(CH_3)CH_3$ (2-methylpropane).

E. Make all the possible isomers of pentane (C_5H_{12}). On the answer sheet write all the partially condensed formulas for the isomers of pentane.

F. Make all the possible isomers of chlorobutane (C_4H_9Cl). Write the partially condensed formulas for the isomers of chlorobutane.

G. Construct two models of methane.

H. In each of the models from *G*, replace one hydrogen with one chlorine. Verify that both models are identical or superimposable.

I. Replace a second hydrogen on both of the models from *H* with a bromine to make bromochloromethane. Again verify that the models are identical.

J. Replace a third hydrogen on both of the models from *I* with a fluorine to make bromochlorofluoromethane. Are the models still superimposable? Be careful before you answer this. The students who obtain superimposable models should raise their hands (author's prediction: about one-half of the class will raise their hands). Those who raise their hands should take <u>one</u> model and switch the hydrogen and chlorine atoms. Now no one in the class should have superimposable models. You should observe that the models are nonsuperimposable mirror images (enantiomers) of each other. This phenomenon most commonly occurs when at least one carbon in the molecule has four different groups attached to it (in this case H, Br, Cl, F). This creates a stereogenic center and in this case, a chiral molecule. The phenomenon of chirality is quite common in biological systems. For example, amino acids are chiral and only one of the two enantiomers of each amino acid is biologically active in animals. Although enantiomers have identical physical properties (melting point, boiling point, density, color), they rotate the plane of polarized light in equal but **opposite** directions. Enantiomers also interact differently with other chiral molecules. Since many molecules in your body are chiral, enantiomers of compounds behave differently in your body. Only one of the two enantiomers of each amino acid is of use to your metabolic system. The compounds in spearmint and caraway (*d* and *l*-carvone) are enantiomers and yet smell differently to most people because the receptors in our noses are chiral.

K. Make a model of ethylene (or ethene, $CH_2 = CH_2$). With most model kits, the model will incorrectly make it appear that the double bond in ethylene consists of two identical bonds. Better model kits show that the double bond consists of a σ and a π bond. Additionally, many kits provide only sp^3 hybridized carbons (4 symmetrically spaced holes 109.5° apart). Thus the model of ethylene will incorrectly show an H-C-H bond angle of 109.5°. Carbon is sp^2 hybridized with 120° bond angles when it has one π bond. Record the correct bond angles. Is rotation possible about the carbon - carbon double bond?

L. Replace a hydrogen on the ethylene with a methyl group to make propene (C_3H_6).

M. There are four different types of hydrogens in propene thus there are four different ways to replace a hydrogen with a methyl group. Make all four butenes and draw the Lewis structure of each.

N. Make all the isomers of chloropropene (C_3H_5Cl). Write the partially condensed formulas for the isomers of chloropropene.

O. Make a model of acetylene (C_2H_2). Record the bond angles.

P. Make a model of cyclohexane (C_6H_{12}). What are the bond angles for a planar regular hexagon? With four groups of electrons about each carbon of cyclohexane, what bond angles do the carbons attempt to achieve? Notice that the model does not assume a planar form but is most stable in either a chair or boat conformation. Cyclohexane molecules spend most of their time in the chair conformation. Give the bond angles for the chair conformation.

Results and Discussion - *Experiment 26*
MOLECULAR MODELS OF ORGANIC COMPOUNDS

Name_____Date_____Lab Section_____

Molecule	Lewis structure	Bond Angles

A. CH_4

_____
H-C-H

B. C_2H_6

_____
H-C-H

_____
H-C-C

C. C_3H_8

_____
H-C-H

_____
H-C-C

_____
C-C-C

D. C_4H_{10}

_____
H-C-H

_____
H-C-C

_____
C-C-C

b.p.   _____   _____

m.p.   _____   _____

Molecule	Partially Condensed Lewis structures	Bond Angles

E. C_5H_{12}

<table>
<tr><td></td><td></td><td>_____
H-C-H</td></tr>
<tr><td></td><td></td><td>_____
H-C-C</td></tr>
<tr><td></td><td></td><td>_____
C-C-C</td></tr>
</table>

F. C_4H_9Cl

<table>
<tr><td></td><td></td><td>_____
H-C-H</td></tr>
<tr><td></td><td></td><td>_____
H-C-C</td></tr>
<tr><td></td><td></td><td>_____
C-C-C</td></tr>
<tr><td></td><td></td><td>_____
H-C-Cl</td></tr>
<tr><td></td><td></td><td>_____
C-C-Cl</td></tr>
</table>

G, H, I. No recorded answers necessary.

J. What is the most common cause of chirality?

Molecule	Lewis structures	Bond Angles

K. C_2H_4

_____
H-C-H

_____
H-C-C

L. C_3H_6

_____
H-C-H

_____
H-C-C

_____
H-C-H

_____
H-C-C

_____
C-C-C

Molecule	Lewis structures

M. C_4H_8

Molecule	Lewis structures

N. C_3H_5Cl

Molecule	Lewis structures	Bond Angles

O. C_2H_2

_____
H-C-C

P. C_6H_{12}

_____
H-C-H

_____
H-C-C

_____
C-C-C

EXERCISES

Exercise 1

SCIENTIFIC NOTATION, SIGNIFICANT FIGURES, ACCURACY AND PRECISION

Select the correct scientific notation representation of each of the following numbers:

	number	a	b	c	d	answer
1.	3.14	3.14	31.4×10^{-1}	0.314×10^1	3.1×10^0	_____
2.	0.08206	8.206×10^2	0.08206	8.21×10^{-2}	8.206×10^{-2}	_____
3.	150.0	1.50×10^2	150	1.500×10^2	1.5×10^2	_____
4.	150	1.50×10^2	150	1.500×10^2	1.5×10^2	_____
5.	0.150	0.150	1.50×10^{-1}	1.5×10^{-1}	1.50×10^1	_____
6.	8314	8314	8.31×10^3	8.314×10^3	83.14×10^2	_____
7.	22.4	22.4	224×10^{-1}	2.24×10^{-1}	2.24×10^1	_____
8.	0.003	0.003	3×10^{-3}	3.00×10^{-3}	3×10^{-2}	_____

Give each of the following in 3 significant figures:

9.	3.14159	3.14	3.15	3.142	none of the answers given	_____
10.	0.08206	0.0820	8.21×10^{-2}	8.21×10^{-1}	none	_____
11.	6.022×10^{23}	6.02×10^{23}	6.03×10^{23}	6.022×10^{23}	none	_____
12.	1.0080	1.00	1	1.01	none	_____
13.	15.994	1.59×10^1	15.9	16.0	none	_____
14.	0.9983	9.98×10^{-1}	9.99×10^{-1}	9.98×10^{-2}	none	_____
15.	35.453	3.54×10^1	3.55×10^1	35.4	none	_____
16.	1.625×10^{-2}	1.62×10^{-2}	1.63×10^{-2}	1.63	none	_____
17.	0.5135	0.513	5.13×10^{-1}	5.14×10^{-1}	none	_____

How many significant figures are in each of the following?

	number	a	b	c	d	answer
18.	24.305	3	4	5	none	_____
19.	260	1	2	3	none	_____
20.	47.90	2	3	4	none	_____
21.	0.0726	3	4	5	none	_____
22.	800	1	2	3	none	_____
23.	14.007	3	4	5	none	_____
24.	0.81	1	2	3	none	_____

Select the best answers to the following calculations. Be sure to use the correct number of significant figures.

	calculation	a	b	c	d	answer
25.	$(12.0)(2.54)$	30.4	3.05×10^1	4.72	none	_____
26.	$0.250/1.057$	2.37×10^{-1}	2.64×10^{-1}	0.250	none	_____
27.	$453.7/16.0$	7.26×10^3	3.53×10^{-2}	2.83×10^1	none	_____
28.	$\dfrac{3.24 \times 10^{22}}{6.023 \times 10^{23}}$	5.38×10^{-1}	5.38×10^{44}	5.38×10^{-2}	none	_____
29.	$1.00/6.02 \times 10^{23}$	6.02×10^{23}	1.66×10^{-24}	1.66×10^{-23}	none	_____
30.	$(6.02 \times 10^{23})(2.5 \times 10^{-4})$	1.5×10^{20}	1.5×10^{19}	1.5×10^{28}	none	_____
31.	$3.2 + 14.19$	17.4	17.39	17.3	none	_____
32.	$3.41 \times 10^4 + 6.27 \times 10^3$	9.68×10^4	4.04×10^4	4.03×10^3	none	_____
33.	$7.2 \times 10^{-4}/8.0 \times 10^{-2}$	9.0×10^{-1}	9.0×10^{-2}	9.0×10^{-3}	none	_____
34.	$4.8 \times 10^3/6.0 \times 10^{-2}$	8.0×10^{-2}	8.0×10^8	8.0×10^2	none	_____
35.	$7.7 \times 10^{-1}/1.1 \times 10^3$	7.0×10^{-4}	7×10^{-4}	7.0×10^{-5}	none	_____
36.	$\dfrac{(5.4 \times 10^2)(8.0 \times 10^{-3})}{(4.0 \times 10^1)(9.0 \times 10^{-5})}$	1.2×10^3	1.2×10^4	1.2×10^2	none	_____
37.	$\dfrac{(3.6 \times 10^{-1})(4.9 \times 10^4)}{(7.0 \times 10^{-2})(6.0 \times 10^{-3})}$	4.2×10^{-3}	4.2×10^8	4.2×10^7	none	_____

The mass of a 5.000 gram piece of iron is measured 3 times on two different balances with the following results:

Trial	balance 1 (grams)	balance 2 (grams)
1	5.01	4.97
2	4.99	4.88
3	5.02	4.91

38. The average for *balance 1* is: a. 4.99 g   b. 5.00 g   c. 5.01 g   _____

39. The average deviation for *balance 1* is: a. 0.01 g   b. 0.04 g   c. 0.02 g   _____

40. The average for *balance 2* is: a. 4.92 g   b. 4.90 g   c. 4.97 g   _____

41. The average deviation for *balance 2* is: a. 0.00 g   b. 0.03 g   c. 0.10 g   _____

42. Which balance is more accurate (based on these measurements)?   a. bal. 1   b. bal. 2   _____

43. Which balance is more precise (based on these measurements)?   a. bal. 1   b. bal. 2   _____

Exercise 2

UNIT CONVERSIONS

	Convert this measurement to →	these units	a	b	c	answer
1.	2.0 m	cm	2.0×10^2 cm	2.0×10^{-2} cm	none	_____
2.	0.250 L	mL	2.50×10^2 mL	2.50×10^{-4} mL	none	_____
3.	65 kg	g	6.5×10^{-2} g	6.5×10^{-4} g	none	_____
4.	5.00 cm	km	5.00×10^{-5} km	5.00×10^5 km	none	_____
5.	3.00 kg	mg	3.00×10^{-6} mg	3.00×10^6 mg	none	_____
6.	68.0 in	cm	173 cm	26.7 cm	none	_____
7.	6.00 qt	L	6.34 L	5.68 L	none	_____
8.	1.50×10^2 lb	g	6.81×10^3 g	3.31×10^{-3} g	none	_____
9.	3.00 g	lb	6.61×10^{-3} lb	3.31×10^3 lb	none	_____
10.	5.00 cm	in	12.7 in	1.97×10^{-1} in	none	_____
11.	325 mL	qt	3.07×10^{-1} qt	3.44×10^{-1} qt	none	_____
12.	5.280×10^3 ft	m	1.760×10^3 m	1.609×10^3 m	none	_____
13.	4.00 oz	g	5.51×10^{-4} g	2.90×10^4 g	none	_____
14.	0.200 ga	mL	757 mL	846 mL	none	_____
15.	2.0 yr	sec	6.3×10^8 sec	1.1×10^7 sec	none	_____
16.	$0.90/ga	$/L	$0.21/L	$0.24/L	none	_____
17.	55.0 miles/hr	km/hr	13.7 km/hr	88.5 km/hr	none	_____
18.	0.500 L	pt	1.06 pt	9.46×10^{-1} pt	none	_____
19.	196 K	°C	-77°C	469°C	none	_____
20.	25°C	K	248 K	298 K	none	_____
21.	98.6°F	°C	37.0°C	23.0°C	none	_____
22.	0.00 K	°F	-524°F	-460°F	none	_____
23.	-40°C	°F	-104°F	10°F	none	_____
24.	70°F	K	294 K	252 K	none	_____

25. 23.9 grams of water at 100°C occupy a volume of 25.0 mL. What is the density of water at 100°C?

   a. 0.998 g/mL   b. 0.956 g/mL   c. 1.05 g/mL   d. none (of the previous answers)   _____

26. If the density of gasoline is 0.70 g/mL, what is the mass in grams of 5.0×10^1 L of gasoline?

   a. 35 g   b. 1.4×10^{-4} g   c. 3.5×10^4 g   d. none   _____

27. The density of acetic acid is 1.05 g/mL. How many mL do you need to provide 125 grams of acetic acid?

   a. 119 mL   b. 131 mL   c. 8.4×10^{-4} mL   d. none   _____

28. A cylinder of titanium (length = 5.00 cm, diameter = 1.20 cm) has a mass of 25.5 g. What is the density of titanium?

   a. 4.51 g/cm^3 b. 1.13 g/cm^3 c. 0.222 g/cm^3 d. none   _____

29. The density of iron is 7.86 g/cm^3. What is the mass of a piece of iron that raises the level of water form the 18.3 mL mark to the 24.8 mL mark when it is placed in a graduated cylinder?

   a. 144 g   b. 1.21 g   c. 51.1 g   d. none   _____

30. The density of lead is 11.3 g/cm^3. What is the volume of 25 g of lead?

   a. 2.2 cm^3 b. 0.45 cm^3 c. 2.8×10^2 cm^3 d. none   _____

31. How many calories does it take to heat 15 g of water from 20°C to 100°C?

   a. 1.5×10^3 cal   b. 1.2×10^3 cal   c. 3.0×10^2 cal   d. none   _____

32. How many joules are needed to heat 5.0 grams of water from 0°C to 20°C?

   a. 1.0×10^2 J   b. 84 J   c. 4.2×10^2 J   d. none   _____

Exercise 3

DENSITY OF WATER VERSUS TEMPERATURE (GRAPH)

On the accompanying piece of graph paper plot the density of water on the vertical axis and the corresponding temperature on the horizontal axis.

temperature (°C)	density (g/mL)	temperature (°C)	density (g/mL)
-20	0.9935	50	0.9881
-10	0.9981	60	0.9832
0	0.9999	70	0.9778
4	1.0000	80	0.9718
10	0.9997	90	0.9653
20	0.9982	100	0.9584
30	0.9957	110	0.9510
40	0.9922		

1. Almost all compounds contract in volume with a decrease in temperature and undergo another contraction upon freezing. The density of ice is 0.9168 g/mL at 0°C. The behavior of water can be described as: normal or anomalous? Explain your answer.

2. What could happen to a sealed container full of water that is cooled until all the water freezes?

3. If water behaved the way most liquids do when they freeze, describe what would happen to lakes during cold winter months and the consequences this would have on life in the lake.

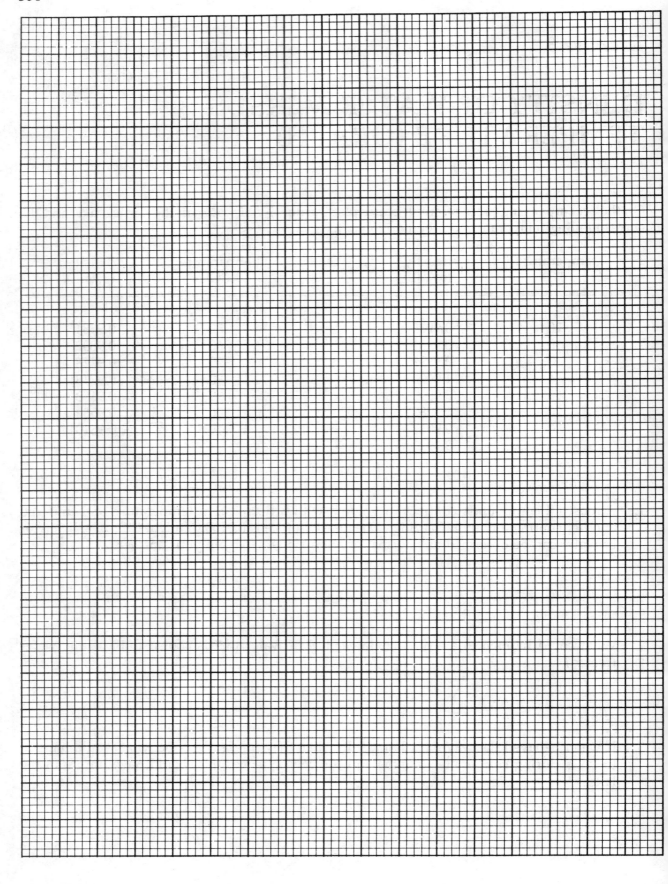

Exercise 4

NAMES AND SYMBOLS OF THE ELEMENTS

This exercise has been designed to familiarize you with the names and symbols of about ½ of the elements. Fill in the name, symbol and multiple choice answer for each problem. The multiple choice part has been designed so it can be answered on a Scantron type form.

	Element	Symbol	Multiple Choice

Choices for 1-5:   a. Br b. Bi c. B d. Ba e. Be

1. Compounds of this group IIIA (or 13) element are important in cleaners (*also what one termite might say to another*)

2. A heavy alkaline earth element. Despite the toxicity of its salts, one of the salts (sulfate) is used in the "milk shakes" taken by a patient for a gastrointestinal series of X-rays (*also what happens to patients when the doctor fails*)

3. The heaviest nonradioactive element

4. This non-metal is the only liquid halogen

5. Compounds of the first member of the alkaline earth metals are extremely toxic

Choices for 6-9:   a. Ge b. Hg c. He d. H

6. A metalloid used extensively in semiconductor material whose name sounds a little like a flower

7. The only liquid metal. Used for thermometers and to make amalgams for items such as tooth fillings (*also a Greek streaker who wore shoes with wings*)

8. The lightest gas. By far the most common element in the sun and universe

9. An inert gas used in balloons (*also what a doctor tries to do to patients*)

306

	Element	Symbol	Multiple Choice

Choices for 10-14:   a. Na   b. Ni   c. Ne   d. N   e. Zn

10. Transition metal used in many alloys such as brass and to galvanize steel (*also where you pour stale milk*)   _____   _____   _____

11. Alkali metal whose ion is blamed for high blood pressure   _____   _____   _____

12. Inert gas used in colored display lights   _____   _____   _____

13. Transition metal whose ion has a green color   _____   _____   _____

14. Nonmetal which is 78% of air (*also what one trogen says to another on the way to bed*)   _____   _____   _____

Choices for 15-19:   a. C   b. Ce   c. Ca   d. Co   e. Cu

15. The first and most abundant rare earth in the Lanthanide series   _____   _____   _____

16. Compounds of this alkaline earth metal make up limestone, chalk, teeth and bones   _____   _____   _____

17. This transition metal is used for electrical wiring   _____   _____   _____

18. This is the transition metal present in Vitamin B_{12} (*also what happens to men as they grow old*)   _____   _____   _____

19. Organic chemistry is the chemistry of compounds of this element. It occurs as diamond and graphite.   _____   _____   _____

Choices for 20-24:   a. I   b. K   c. Kr   d. Mg   e. Mn

20. This inert gas is also the name for Superman's planet   _____   _____   _____

21. Compounds of this alkali metal are sometimes used by people on low sodium diets   _____   _____   _____

22. This transition metal is used in many important alloys including steel   _____   _____   _____

23. This alkaline earth metal is used in fireworks and flash bulbs and is present in chlorophyll   _____   _____   _____

24. This halogen is used as an antiseptic   _____   _____   _____

	Element	Symbol	*Multiple Choice*

Choices for 25-29: a. Si b. S c. Se d. Sn e. Ra

25. The second most abundant element in the earth's crust is used in semiconductors (*also a joker in jail*) _____ _____ _____

26. This alkaline earth metal is radioactive and has been used on watch and clock hands (*also what police do to wild parties*) _____ _____ _____

27. This element is similar to sulfur and is used in photocells and semiconductors _____ _____ _____

28. This nonmetal is used in the vulcanization of natural rubber and as a fungicide _____ _____ _____

29. This member of the carbon family has predominantly metallic properties and is used to coat steel in cans for preserving food _____ _____ _____

Choices for 30-34: a. P b. Pt c. Pb d. Pu e. U

30. This transuranium element is used in nuclear weapons and reactors and is extremely toxic _____ _____ _____

31. This silvery white, rare transition metal is used in jewelry and for electrical contacts (*also what you might like to do to a big meanie*) _____ _____ _____

32. This nonmetal is used in the manufacture of match heads, pesticides and detergent additives _____ _____ _____

33. This radioactive element is used in nuclear reactors and weapons _____ _____ _____

34. This metal is used in the manufacture of storage batteries, gasoline and X-ray shields _____ _____ _____

308

	Element	Symbol	Multiple Choice

Choices for 35-39:   a. Ar   b. Al   c. Au   d. Ag   e. As

35. This inert gas is the third most abundant gas in the atmosphere   _____   ____   ____

36. This member of Group IIIA (or 13) is used extensively in foil, kitchen utensils and wiring   _____   ____   ____

37. This member of the nitrogen family is used in the manufacture of pesticides   _____   ____   ____

38. This transition metal has often been used as the base for monetary systems and is used in jewelry   _____   ____   ____

39. This brilliant white metal is an excellent electrical conductor and used in jewelry and tooth fillings   _____   ____   ____

Choices for 40-43:   a. O   b. W   c. Ti   d. Li

40. Compounds of this member of the alkali metals are used to treat some types of mental disorders   _____   ____   ____

41. This nonmetal is the second most abundant element in the atmosphere and the most abundant in the earth's crust   _____   ____   ____

42. This transition metal is used for light bulb filaments   _____   ____   ____

43. This transition metal is as strong as steel but 45% less dense   _____   ____   ____

Choices for 44-47:   a. F   b. Fe   c. Cl   d. Cr

44. Compounds of this halogen are used as additives in toothpastes *(also the opposite of roofing)*   _____   ____   ____

45. This halogen is used in bleaches and pools   _____   ____   ____

46. This transition metal is used as a coating on water taps and bumpers because of its luster   _____   ____   ____

47. This transition metal is the main element in steel *(also a really pressing thing)*   _____   ____   ____

Exercise 5

PROTONS, NEUTRONS, ELECTRONS AND ISOTOPES

Choose the correct number of protons, electrons and neutrons (for the examples given, there is only one abundant natural isotope).

element or ion	a	b	c	d	answer
1. H	1, 1, 0	1, 1, 1	1, 0, 0	none	_____
2. He	4, 4, 4	2, 2, 2	2, 2, 4	none	_____
3. F	10, 10, 9	9, 9, 10	9, 9, 9	none	_____
4. F^-	10, 9, 10	9, 10, 10	9, 10, 9	none	_____
5. Na	11, 12, 12	11, 10, 12	12, 12, 11	none	_____
6. Na^+	11, 10, 12	11, 13, 12	12, 11, 11	none	_____
7. Mn	55, 55, 25	25, 25, 30	30, 30, 25	none	_____
8. Mn^{2+}	25, 27, 30	30, 28, 25	25, 23, 30	none	_____
9. I	74, 74, 53	53, 54, 74	53, 53, 74	none	_____
10. I^-	74, 73, 53	53, 54, 74	53, 53, 74	none	_____
11. Bi	83, 83, 126	126, 126, 83	83, 80, 126	none	_____
12. Bi^{3+}	83, 83, 126	83, 80, 126	120, 123, 83	none	_____

13. On February 29, 2002, a Modesto Junior College chemistry class will discover a new element, Modestonium (Mm). The element will have a colorful electron shell and a tasty nucleus. Mass spectrometric analysis will reveal that the element consists of 80.00% $^{296.0}_{114}Mm$ and 20.00% $^{294.0}_{114}M$. What atomic mass will appear for Mm in the periodic table?

a. 295.8   b. 294.4   c. 295.6   d. none of the previous answers   _____

14. $^{235}_{92}U$ undergoes nuclear fission when impacted by neutrons and is therefore utilized as a fuel in nuclear reactors. $^{238}_{92}U$ does not undergo nuclear fission when impacted by neutrons (although it can be converted to fissionable $^{239}_{94}Pu$ in "breeder reactors"). Because of the natural abundances of the uranium isotopes (234 - 0.0057%, 235 - 0.72%, 238 - 99.27%), fuel for nuclear reactors is expensive (uranium is a relatively rare element and it is difficult to separate isotopes). How many neutrons are in $^{235}_{92}U$ and $^{238}_{92}U$ respectively?

   a. 235, 238   b. 143, 146   c. 92, 92   d. none of the previous answers   _____

15. An alternative to nuclear fission reactors that is still in the development stage involves the use of nuclear fusion reactions that utilize one of the isotopes of hydrogen, $^{2}_{1}H$ (often called deuterium). There are three known isotopes of hydrogen, two of which are naturally occurring (99.985% $^{1}_{1}H$, 0.015% $^{2}_{1}H$). The third isotope is radioactive ($^{3}_{1}H$ - tritium) and has a half-life of 12.26 years. Although the percentage of deuterium in nature is very low, there is sufficient hydrogen around to supply fuel for deuterium fusion reactors for millions of years. How many neutrons are in the three isotopes of hydrogen ($^{1}_{1}H$, $^{2}_{1}H$, $^{3}_{1}H$) respectively?

   a. 1, 2, 3   b. 1, 1, 1   c. 0, 0, 0   d. 0, 1, 2   e. none   _____

16. Which of the following would probably be a stable isotope of aluminum?

   a. $^{30}_{13}Al$ b. $^{27}_{13}Al$ c. $^{24}_{13}Al$ d. all should be stable   _____

17. Which of the following would probably be a radioactive isotope of oxygen?

   a. $^{16}_{8}O$ b. $^{13}_{8}O$ c. $^{14}_{7}N$ d. none of the previous answers   _____

Questions 18-20 are based on the information below. The actual masses and abundances of the isotopes of magnesium, copper and lead are given. Calculate the average atomic mass for each element.

element	isotope mass (amu)	abundance	element	isotope mass (amu)	abundance
magnesium	23.985	78.70	lead	203.973	1.48
	24.986	10.13		205.974	23.6
	25.982	11.17		206.976	22.6
				207.977	52.3
copper	62.930	69.09			
	64.928	30.91			

18. Magnesium: a. 24.31   b. 25.00   c. 24.986   d. none   _____

19. Copper: a. 64.00   b. 62.93   c. 63.55   d. none   _____

20. Lead: a. 206.2   b. 207.2   c. 207.5   d. none   _____

Exercise 6

POLYATOMIC IONS

A. Questions on nomenclature rules

1. The negative ions of the elements end with the letters

    a. ite   b. ate   c. ide   d. ine   e. none   _____

2. Chlorite, nitrite and sulfite

    a. have one more oxygen than chlorate, nitrate and sulfate
    b. have one less oxygen than chlorate, nitrate and sulfate
    c. are the negative ions of the elements
    d. all have two oxygens
    e. all have three oxygens   _____

3. For aqueous solutions of the following compounds, the hydrogen is
    dropped and the ate ending is:

    hydrogen acetate
    hydrogen carbonate
    hydrogen chlorate
    hydrogen iodate
    hydrogen nitrate
    hydrogen oxalate
    hydrogen phosphate
    hydrogen sulfate

    a. changed to ous acid (along with spelling changes in some cases)
    b. changed to ic acid (along with spelling changes in some cases)
    c. left unchanged
    d. none of the above   _____

4. For thiosulfate and thiocyanate, the prefix thio means

    a. three
    b. the ion has a negative 3 charge
    c. an oxygen has been replaced by a sulfur
    d. none of the above   _____

B. For the polyatomic ions listed below, first give the charges on the ion and then go back and give the number of oxygens in the ion.

	a.	0			a.	0
	b.	-1			b.	1
	c.	-2			c.	2
	d.	-3			d.	3
	e.	none			e.	4

ion	problem number	charge	problem number	number of oxygens
acetate	1.	_____	26.	_____
bromate	2.	_____	27.	_____
bromide	3.	_____	28.	_____
carbonate	4.	_____	29.	_____
chlorate	5.	_____	30.	_____
chloride	6.	_____	31.	_____
chlorite	7.	_____	32.	_____
chromate	8.	_____	33.	_____
cyanide	9.	_____	34.	_____
fluoride	10.	_____	35.	_____
hydrogen carbonate	11.	_____	36.	_____
hydroxide	12.	_____	37.	_____
hypochlorite	13.	_____	38.	_____
iodate	14.	_____	39.	_____
iodide	15.	_____	40.	_____
nitrate	16.	_____	41.	_____
nitrite	17.	_____	42.	_____
oxalate	18.	_____	43.	_____
oxide	19.	_____	44.	_____
perchlorate	20.	_____	45.	_____
permanganate	21.	_____	46.	_____
phosphate	22.	_____	47.	_____
sulfate	23.	_____	48.	_____
sulfide	24.	_____	49.	_____
sulfite	25.	_____	50.	_____

C. Match the name of each negative ion with the correct formula. Fill in the appropriate name or code on this sheet and/or darken all the letters on a Scantron answer sheet.

ion	code		problem number	ion	answer
Br^-	a		1.	acetate	_____
BrO_3^-	b		2.	bicarbonate	_____
CO_3^{2-}	c		3.	bisulfate	_____
$C_2H_3O_2^-$	d		4.	bisulfite	_____
$C_2O_4^{2-}$	e		5.	bromate	_____
CN^-	ab		6.	bromide	_____
Cl^-	ac		7.	carbonate	_____
ClO^-	ad		8.	chlorate	_____
ClO_2^-	ae		9.	chloride	_____
ClO_3^-	bc		10.	chlorite	_____
ClO_4^-	bd		11.	chromate	_____
CrO_4^{2-}	be		12.	cyanide	_____
$Cr_2O_7^{2-}$	cd		13.	dichromate	_____
F^-	ce		14.	fluoride	_____
HCO_3^-	de		15.	hydroxide	_____
HSO_3^-	abc		16.	hypochlorite	_____
HSO_4^-	abd		17.	iodate	_____
I^-	abe		18.	iodide	_____
IO_3^-	acd		19.	nitrate	_____
MnO_4^-	ace		20.	nitrite	_____
NO_2^-	ade		21.	oxalate	_____
NO_3^-	bcd		22.	oxide	_____
O^{2-}	bce		23.	perchlorate	_____
OH^-	bde		24.	permanganate	_____
PO_4^{3-}	cde		25.	phosphate	_____
S^{2-}	abcd		26.	sulfate	_____
SCN^-	abce		27.	sulfide	_____
SO_3^{2-}	abde		28.	sulfite	_____
SO_4^{2-}	acde		29.	thiocyanate	_____
$S_2O_3^{2-}$	bcde		30.	thiosulfate	_____

Exercise 7

FORMULAS OF COMPOUNDS

Select the correct formula for each name given.

	name	a	b	c	d	answer
1.	barium nitrate	$Ba(NO_3)_2$	$BaNO_3$	Ba_3N_2	none	_____
2.	phosphoric acid	H_3PO_3	H_2PO_4	H_3PO_4	none	_____
3.	potassium iodate	K_2IO_3	KIO_3	KIO_4	none	_____
4.	sodium dichromate	$Na(CrO_4)_4$	Na_2CrO_4	$Na_2Cr_2O_7$	none	_____
5.	iron(III) bromide	$FeBr_3$	Fe_2Br_3	$FeBr$	none	_____
6.	cupric phosphate	Cu_3P_2	$Cu_2(PO_4)_3$	$CuPO_4$	none	_____
7.	carbonic acid	$H_2C_2O_4$	$HC_2H_3O_2$	H_2CO_3	none	_____
8.	calcium oxalate	CaC_2O_4	$Ca(C_2O_4)_2$	$CaCO_3$	none	_____
9.	zinc sulfate	ZnS	$ZnSO_4$	$ZnSO_3$	none	_____
10.	barium hydroxide	$BaOH$	$Ba(OH)_3$	$Ba(OH)_2$	none	_____
11.	potassium chlorate	KCl	$KClO_3$	$KClO_2$	none	_____
12.	tin(II) sulfide	SnS	TiS	SnS_4	none	_____
13.	ferrous chloride	$FeCl_3$	$FeClO_3$	$FeClO_2$	none	_____
14.	sulfuric acid	H_2SO_4	H_2S	H_2SO_3	none	_____
15.	magnesium carbonate	$MnCO_3$	$Mg(HCO_3)_2$	$MgCO_3$	none	_____
16.	silver chromate	Si_2CrO_4	Ag_2CrO_4	$Hg_2Cr_2O_7$	none	_____
17.	ammonium nitrate	NH_4NO_3	NH_3NO_2	NH_4NO_2	none	_____
18.	mercury(II) oxide	Hg_2O	HgO	HgO_2	none	_____
19.	aluminum sulfate	$AlSO_4$	Al_2SO_4	$Al_2(SO_4)_3$	none	_____
20.	stannic bromide	$SnBr_4$	$SnBr_3$	$SnBr_2$	none	_____

	name	a	b	c	d	answer
21.	nitric acid	H_3N	HNO_3	HNO_2	none	_____
22.	strontium nitrate	$SrNO_3$	$Sr(NO_2)_2$	$Sr(NO_3)_2$	none	_____
23.	lead(II) sulfide	PbS	Pb_2S	$PbSO_4$	none	_____
24.	potassium oxide	K_2O	KO	KO_2	none	_____
25.	nitrogen dioxide	NO_2^-	NO_2	N_2O	none	_____
26.	sodium bicarbonate	$Na(CO_3)_2$	Na_2CO_3	$NaHCO_3$	none	_____
27.	hydrochloric acid	$HClO_3$	HCl	$HClO_2$	none	_____
28.	ammonium sulfide	NH_4S	NH_4S_2	$(NH_3)_2S$	none	_____
29.	cobalt(II) iodide	CoI_2	Co_2I	$Co(IO_3)_2$	none	_____
30.	cerium(III) hydroxide	Ce_3OH	$Ce(OH)_3$	$Ce_2(OH)_3$	none	_____

B. Give the formulas of the following compounds:

1. zinc phosphate   _____
2. silver nitrate   _____
3. potassium chromate   _____
4. tin(IV) oxide   _____
5. iron(III) sulfate   _____
6. nitrous acid   _____
7. copper(I) sulfide   _____
8. magnesium bromide   _____
9. lithium phosphate   _____
10. calcium carbonate   _____
11. hydrobromic acid   _____
12. sulfur dioxide   _____
13. copper(II) acetate   _____
14. sodium nitrite   _____
15. ammonium chloride   _____

Exercise 8

NOMENCLATURE

A. Select the correct name for each formula given.

	formula	a	b	c	d	answer
1.	KBr	phosphorous bromide	potassium bromate	potassium bromide	none	_____
2.	NH_3	ammonia	ammonium hydroxide	ammonium	none	_____
3.	HNO_3	hydrogen nitrite	nitric acid	nitrous acid	none	_____
4.	$BaSO_4$	barium sulfate	barium sulfide	barium sulfite	none	_____
5.	Cu_2O	dicopper oxide	copper(II) oxide	cupric oxide	none	_____
6.	$Fe_2(SO_4)_3$	ferrous sulfate	iron(III) sulfate	iron(III) sulfide	none	_____
7.	N_2O_4	nitrogen oxide	nitrogen tetroxide	dinitrogen tetroxide	none	_____
8.	K_3PO_4	potassium phosphate	potassium phosphide	calcium phosphate	none	_____
9.	Al_2O_3	ammonium oxide	aluminum oxide	aluminate	none	_____
10.	NH_4Cl	ammonium chloride	ammonia chloride	nitrogen hydrogen chloride	none	_____
11.	$Cu(OH)_2$	copper hydroxide	copper dihydroxide	copper(II) hydroxide	none	_____
12.	SrS	silver sulfate	strontium sulfide	strontium sulfate	none	_____
13.	Ag_2O	silver oxide	disilver oxide	mercury(I) oxide	none	_____
14.	Na_2CrO_4	sodium dichromate	sodium chromate	nitrogen chromate	none	_____
15.	$Ca(C_2H_3O_2)_2$	calcium oxalate	carbon acetate	calcium acetate	none	_____
16.	H_2SO_4	sulfurous acid	hydrogen sulfite	hydrogen sulfide	none	_____
17.	NaOH	sodium hydroxide	sodium oxyhydride	neon hydroxide	none	_____
18.	$ZnSO_4$	zirconium sulfate	zinc sulfide	zinc sulfate	none	_____
19.	SiO_2	silicon oxide	silicon dioxide	silver oxide	none	_____
20.	KI	phosphorous iodide	potassium iron	potassium iodide	none	_____
21.	$HC_2H_3O_2$	oxalic acid	carbonic acid	acetic acid	none	_____
22.	Na_2CO_3	sodium carbonate	sodium bicarbonate	sodium oxalate	none	_____

317

	formula	a	b	c	d	answer
23.	HCl	hydrogen chlorate	hydrogen chlorite	hydrochloric acid	none	_____
24.	SnO_2	tin(IV) oxide	tin dioxide	tin(II) oxide	none	_____
25.	MgC_2O_4	manganese oxalate	magnesium acetate	magnesium oxalate	none	_____
26.	CuS	copper(I) sulfide	copper(II) sulfide	copper(II) sulfate	none	_____
27.	$SnCl_2$	tin dichloride	tin(IV) chloride	tin(II) chloride	none	_____
28.	$Zn(CN)_2$	zinc cyanide	zinc carbonate	zinc dicyanide	none	_____
29.	$KHCO_3$	potassium dicarbonate	potassium bicarbonate	potassium acetate	none	_____
30.	SO_3	sulfite	sulfate	sulfur trioxide	none	_____

B. Name the following compounds:

1. $Ca(NO_3)_2$ _____

2. Na_2SO_3 _____

3. $HClO_{3(aq)}$ _____

4. $Hg(NO_3)_2$ _____

5. $(NH_4)_2CO_3$ _____

6. LiClO   _____

7. $NiSO_4$ _____

8. $H_2SO_{4(aq)}$ _____

9. N_2O _____

10. $KMnO_4$ _____

11. $Zn(C_2H_3O_2)_2$ _____

12. Fe_2S_3 _____

13. $Cu(OH)_2$ _____

14. $BaCl_2$ _____

15. $CePO_4$ _____

Exercise 9

ELECTROMAGNETIC RADIATION: ENERGY VERSUS FREQUENCY AND WAVELENGTH (GRAPHS)

The table below gives the energy of light and its corresponding wavelengths and frequencies. Plot on the same graph the energy (y axis) versus the wavelength (x axis) and energy (y axis) versus the frequency (x axis).

Energy (joules/mol) ($\times 10^{-5}$)	wavelength (nanometers)	frequency (cps)
4.78	2.50×10^2	1.20×10^{15}
3.98	3.00×10^2	1.00×10^{15}
2.99	4.00×10^2	7.50×10^{14}
2.39	5.00×10^2	6.00×10^{14}
1.99	6.00×10^2	5.00×10^{14}
1.71	7.00×10^2	4.28×10^{14}
1.49	8.00×10^2	3.75×10^{14}
1.33	9.00×10^2	3.33×10^{14}

1. Based on the graph, is the energy directly proportional to the wavelength or frequency? Explain your answer.

2. Does the energy increase or decrease as the wavelength increases?

3. The visible region of the electromagnetic radiation spectrum extends from about 400 nm to 700 nm. In what spectral regions are the wavelengths 250 nm and 900 nm respectively?

4. The carbon-carbon bond energy in acetone, CH_3-CO-CH_3 is 339 kJ/mol. What wavelength is potentially enough energy to break an acetone C-C bond?

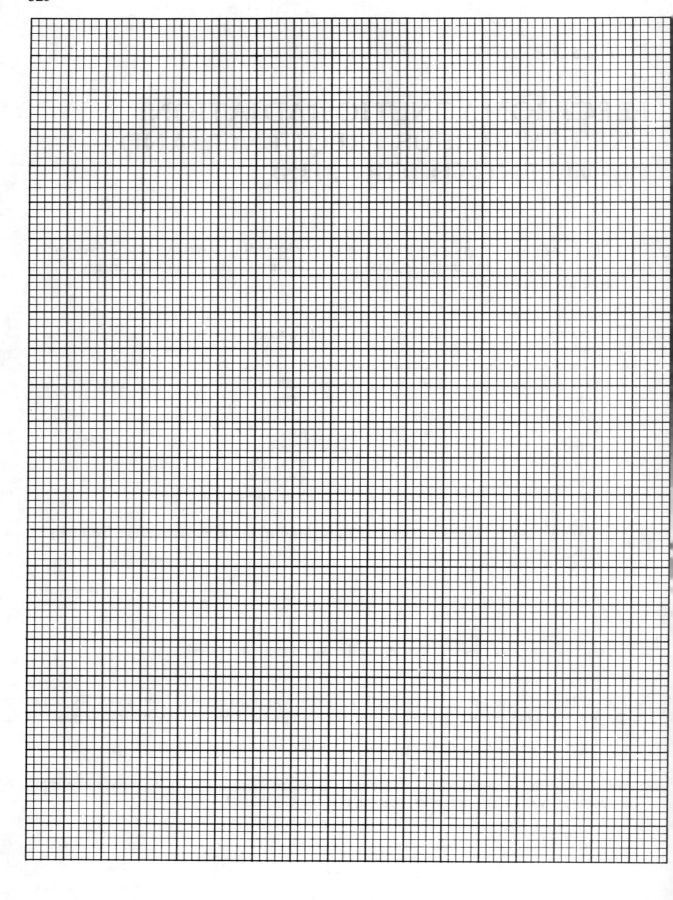

Exercise 10

ELECTRONIC STRUCTURES AND LEWIS STRUCTURES OF ATOMS

The following are the multiple choice answers for numbers 1-14. Write in the correct answer or darken the correct answer on the Scantron. When double letters appear, darken both letters.

a. $1s^2 2s^2 2p^3$

b. $1s^2 2s^5$

c. $1s^2 2s^2 2p^6$

d. $1s^2 2s^2 2p^6 3s^2 3p^1$

e. $1s^2 2s^2 2p^6 3s^2 3p^2$

ab. $1s^2 2s^2 2p^6 3s^2 3p^4$

ac. $1s^2 2s^2 2p^6 3s^2 3p^5$

ad. $1s^2 2s^2 2p^6 3s^2 3p^6$

ae. $1s^2 2s^2 2p^6 3s^2 3p^6 4s^1$

bc. $1s^2 2s^2 2p^6 3s^2 3p^6 4s^2$

bd. $1s^2 2s^2 2p^6 3s^2 3p^6 3d^2$

be. $1s^2 2s^2 2p^6 3s^2 3p^6 3d^7$

cd. $1s^2 2s^2 2p^6 3s^2 3p^6 4s^2 3p^5$

ce. $1s^2 2s^2 2p^6 3s^2 3p^6 4s^2 3d^5$

de. $1s^2 2s^2 2p^6 3s^2 3p^6 4s^2 3d^{10} 4p^6 5s^2$

1.   Ar   _____

2.   Al   _____

3.   S   _____

4. Ca^{2+} _____

5. S^{2-} _____

6.   N   _____

7. Al^{3+} _____

8.   K   _____

9.   Cl   _____

10. Cl^- _____

11. Mn   _____

12. Sr   _____

13. K^+ _____

14. Ca   _____

15. Which of the above (a - de) are not electronic structures of any element or ion in its ground state?

     a. b   b. bd   c. be   d. cd   e. all of the answers (b, bd, be, cd)   _____

The following valence electron configurations are the multiple choice answers for numbers 16-22. Fill in the correct answer or darken the correct answer on a Scantron form. When double letters appear, darken both letters.

a. s^1

b. s^2

c. s^2p^1

d. s^2p^2

e. s^2p^3

ab. s^2p^4

ac. s^2p^5

ad. s^2p^6

16. Mg _____

17. F   _____

18. O   _____

19. S   _____

20. P   _____

21. Na _____

22. B   _____

For the element indicated, give the multiple choice answer of the correct Lewis structure.

element	a	b	c	d	answer
23. C	·Ċ:	·Ċ:	Ċ:	none	_____
24. Br	Br·	:B̈r:	:B̈r:	none	_____
25. K	K·	K	K̇:	none	_____
26. Al	Al·	·Al·	Al:	none	_____
27. O	:Ö:	·Ö:	O	none	_____
28. N	·N:	·N:	N:	none	_____

For 29-31, give possible values for the quantum numbers n, l, m, s respectively for:

29. a 2s electron: a. 2,1,0,½ b. 2,0,0,½ c. 2,1,1,½ d. none _____

30. a 3d electron: a. 3,3,0,½ b. 3,1,2,½ c. 3,2,2,½ d. none _____

31. a 3p electron: a. 3,3,0,½ b. 3,1,1,½ c. 3,2,1,½ d. none _____

Optional: Write the electron configurations for the following:

32. Si _____

33. As _____

34. O^{2-} _____

35. Zn _____

36. Sr^{2+} _____

PERIODIC PROPERTIES OF THE ELEMENTS (GRAPHS, COMPUTER OPTION)

Graphs of parameters including the atomic radius, ionization energy and electronegativity can be obtained using the table below (**Part A**) or the computer exercise (**Part B**) that follows.

Part A. The table below gives the atomic radius, ionization energy and electronegativity of the first 37 elements. On three separate pieces of graph paper, plot the atomic radius, ionization energy or electronegativity versus the atomic number and look for trends or periodic properties. In these types of graphs, it is appropriate to plot the data point to point.

element	atomic number	atomic radius (nanometers)	ionization energy (kcal/mol)	electronegativity
H	1	0.037	313	2.1
He	2	0.05	567	
Li	3	0.152	124	1.0
Be	4	0.111	215	1.5
B	5	0.088	191	2.0
C	6	0.077	260	2.5
N	7	0.070	336	3.0
O	8	0.066	314	3.5
F	9	0.064	402	4.0
Ne	10	0.070	497	
Na	11	0.186	119	0.9
Mg	12	0.160	176	1.2
Al	13	0.143	138	1.5
Si	14	0.117	188	1.8
P	15	0.110	254	2.1
S	16	0.104	239	2.5
Cl	17	0.099	300	3.0
Ar	18	0.094	363	
K	19	0.231	100	0.8
Ca	20	0.197	141	1.0
Sc	21	0.160	151	1.3
Ti	22	0.146	158	1.5
V	23	0.131	156	1.6
Cr	24	0.125	156	1.6
Mn	25	0.131	171	1.5
Fe	26	0.126	182	1.8
Co	27	0.125	181	1.8
Ni	28	0.124	176	1.8
Cu	29	0.128	178	1.9
Zn	30	0.133	216	1.6
Ga	31	0.122	138	1.6
Ge	32	0.122	187	1.8
As	33	0.121	231	2.0
Se	34	0.117	225	2.4
Br	35	0.114	273	2.8
Kr	36	0.109	323	
Rb	37	0.244	96	0.8

Questions on Periodic Properties

1. Briefly explain the meaning of ionization energy.

2. How does the atomic radius vary as the atomic number increases going across a period? Give reasons for the trend.

3. How does the atomic radius vary as the atomic number increases going down the IA (1) or VIIA (17) groups? Give reasons for the trend.

4. How does the ionization energy vary as the atomic number increases going across a period? Give reasons for the trend.

5. How does the ionization energy vary as the atomic number increases going down the IA (1) or VIIA (17) groups? Give reasons for the trend.

6. Are there any correlations among the variations of the atomic radius, ionization energy and/or electronegativities? Explain your answer.

7. (bonus question) Why does the ionization energy drop going from the IIA (2) to the IIIA (13) group?

8. (bonus question) Why does the ionization energy drop from the VA (15) to the VIA (16) group?

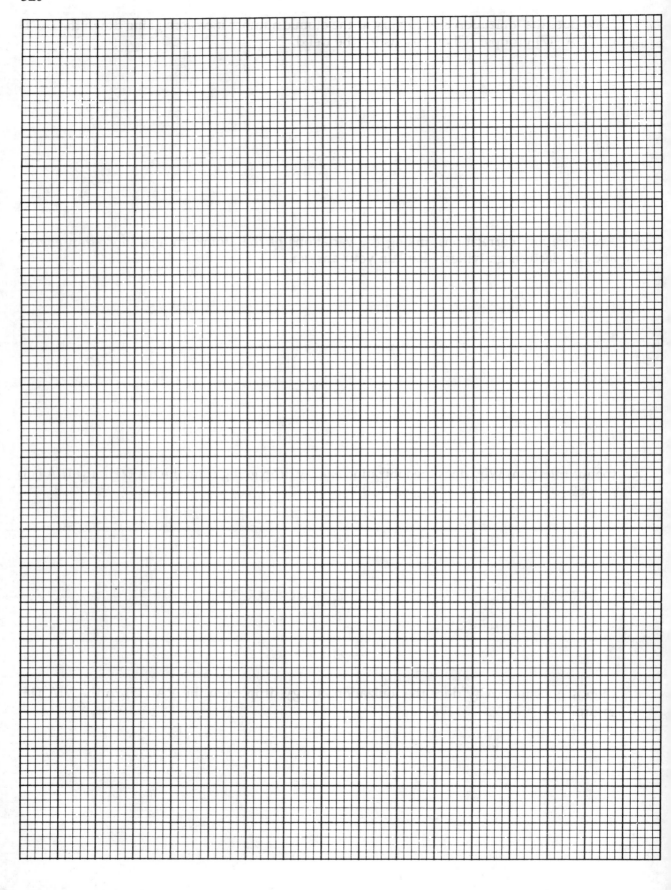

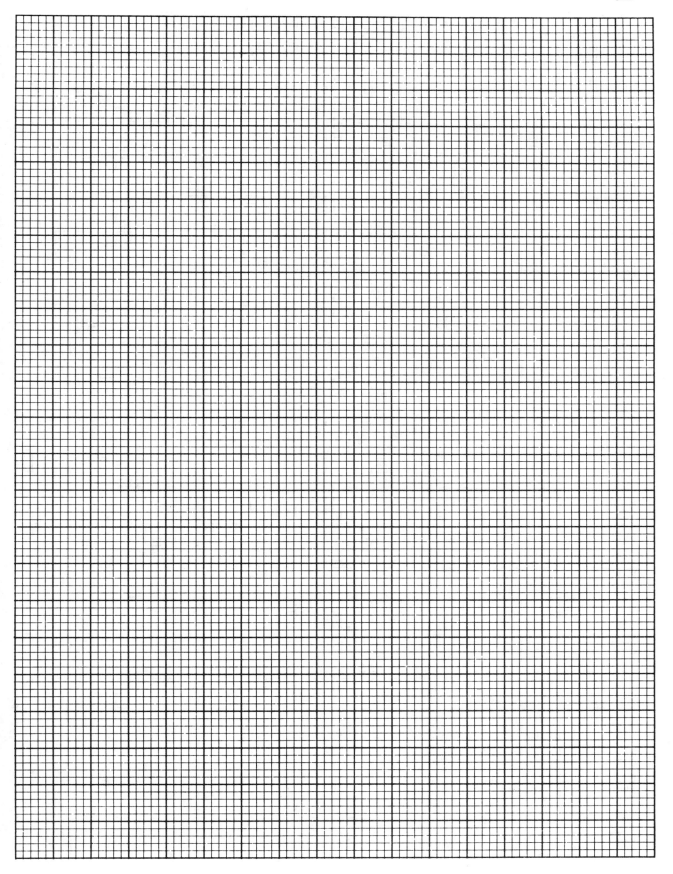

Part B. The software program, "*KC? Discoverer*", available from **Project Seraphim**, can be used to quickly graph many properties of the elements. For example, in this lesson you will graph the ionization energy of the elements versus atomic number. This graph illustrates how ionization energy varies as the elements go across a period or down a group of the periodic table. You will also use this program to locate elements that have properties within certain ranges.

Procedure

Obtain a copy of the software program "*KC? Discoverer*" (for the IBM PC computer) from your instructor or the computer laboratory facility manager in your school's computer laboratory. After loading DOS, on an A prompt (A>), type "graphics" and **Enter** (hit the Enter key). Insert the "*KC? Discoverer*" system disk in Drive A and the program disk in Drive B and type "KC". Follow the instructions until you come to the list of options. Anytime that you want to return to the main menu, press **Escape**.

A. Move the cursor to *Graph* and **Enter**. In a list of attributes and properties, the cursor will appear at the words "atomic number". **Enter**. Next choose *no* for a choice of 2 properties. Now move the cursor with the arrow keys to 1st ionization energy and **Enter**. A graph of ionization energy vs atomic number should appear. Select the enlarge option and choose an atomic number range of 1 - 18 and an appropriate range for ionization energy (base your choice of range on the first graph). **Enter** and after the graph appears, print the graph by pressing "**shift-print screen**". Turn in this graph with your report.

B. Return to the menu. Using the "*Find*" option, produce on the screen a periodic table with elements selected for boiling points between 4 K and 90 K. List the elements that fall in this category on the answer sheet. Also using the instructions below, find the elements with the following properties and list them on the answer sheet (Hint: there should be 7 or less):

menu option	property	range
Find	electrical conductivity	>300 mho-cm (put down a range of 300 to a large value such as 1000 mho-cm)
Find	density	>20 g/cm^3
Find	melting point	>3000 K
Find	melting point	283 K - 323 K
Find	carcinogenicity	positive
Find	reaction: water	vigorous
Sort	cosmic abundance	the 5 most abundant
Sort	earth crust abundance	the 5 most abundant
Sort	cost (pure)	the 5 most expensive

C. Element screens. If you want information about one or more elements, the following instructions apply. For this exercise, we will investigate titanium. Be sure to read the questions in the *Answers* section as the answers to these questions will appear on the screen as you do this exercise. On the main menu, select *Find*. Move the cursor to symbol and **Enter**. Press the **Pg Up** key to reach the second screen and highlight Ti, titanium and **Enter** and **End**. Titanium should be highlighted on the periodic table that appears. Type **E** and the first of three element screens for titanium should appear (what color is it?). Press **Pg Dn** (what is its electronegativity?). Again press **Pg Dn** (describe a titanium crystal?). Press **Esc** to return to the main menu.

D. The *Table* function enables you to graph a property of the elements going across a period or down a group. Unless you desire to do so, do not print out the graphs you will obtain on the screen but based on the screen plots, answer the questions in the *Answers* section.

On the main menu, select *Table* and **Enter**. Upon display of the periodic table, type **G** for graphing. Properties that can be graphed will be displayed. Select atomic radius and then select **P** to indicate that you want to plot radius vs period. The periodic table will reappear. Use the arrow keys to move the cursor to the 2nd period (Li first member of period) and **Enter**. After answering the question in the *Answers* section, **Enter** and **Esc**. Repeat the above process for the following parameters:

property	vs	elements
2. atomic radius		group IA
3. 1st ionization energy		period 2
4. 1st ionization energy		group IA
5. number of isotopes		period 4

E. Consider the criteria you would use to select the material to use for:

1. a bicycle frame
2. electrical wiring in a house
3. pots and pans
4. water pipes in a house
5. a friendship ring

While the best material might be a compound or a mixture, assume <u>for this exercise</u> that you are restricted to the use of a pure element. Use *KC? Discoverer* to compare pertinent properties of possible elements for each of the above applications. List the criteria you use and your selected element.

Answers

A. Graph of ionization energy (atomic numbers 1 - 18) - attach!

B. Fill in the following table:

property	*range*	*elements*
boiling point	4 - 90 K	_____
electrical conductivity	>300 mho-cm	_____
density	>20 g/cm³	_____
melting point	>3000 K	_____
melting point	283 K - 323 K	_____
carcinogenicity	positive	_____
reaction: water	vigorous	_____
cosmic abundance	5 most	_____
earth crust abundance	5 most	_____
cost (pure)	5 most expensive	_____

C. Titanium -   color   _____

electronegativity   _____

crystal type   _____

D. Give and <u>explain</u> the trend for each graph below.

property	*vs*	*elements*
1. atomic radius		period 2

2.   atomic radius   group IA

property	_vs_	_elements_
3. 1st ionization energy		period 2

4. 1st ionization energy   group IA

5. number of isotopes   period 4

E. For each application, give the criteria you used for the selection of the best element for the construction material.

Application	Criteria	Recommended Element
1. a bicycle frame		
2. electrical wiring in a house		
3. pots and pans		
4. water pipes in a house		
5. a friendship ring		

Exercise 12

POLARITIES OF BONDS AND MOLECULES

Bonds between partners with electronegativity differences greater than 1.7 are usually called ionic bonds. For differences between 0.5 and 1.7, the bonds are called polar covalent and bonds with differences of 0.5 or less are usually called nonpolar covalent or covalent. While these values serve as general guidelines, often the environment of the bond in the molecule can have a significant influence on the polarity of the bond. An even simpler way of assigning labels to bonds is to assume that metal-nonmetal and metal-polyatomic bonds are ionic and that nonmetal-nonmetal bonds, with two qualifications are polar covalent. The first qualification is that bonds between two atoms of the same element are nonpolar covalent. The second is that the carbon-hydrogen bond, one of the two most common bonds in organic compounds, experimentally exhibits nonpolar properties and will be considered nonpolar covalent.

Use the information above or the electronegativity data on page 323 to classify the bonds in the following molecules as:

a. ionic

b. polar covalent

c. nonpolar covalent

1. Br_2 _____

2. NO _____

3. HCl _____

4. HBr _____

5. MgO _____

6. CH_4 _____

7. KF _____

8. CCl_4 _____

9. NH_3 _____

10. H_2O _____

11. Na_2O _____

12. N_2 _____

13. CO _____

14. ICl _____

15. NaBr _____

334

For polyatomic molecules, the net polarity of the molecule is determined not only by bond polarities but also by molecular geometry. Although CO_2 has polar covalent bonds, the molecule is nonpolar due to a cancellation of bond polarities. Water on the other hand, is polar as indicated by the arrow because its bent shape enables the vertical portion of the polarity to add instead of cancel.

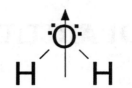

Describe the following molecules as nonpolar (a) or polar (b).

1
carbon tetrachloride

2
chloroform

3   :S̈=C=S̈:
carbon sulfide

4   :S̈—H
hydrogen sulfide

5
ammonia

6   H—C≡N:
hydrogen cyanide

7
formaldehyde

8
ethylene

1. _____

2. _____

3. _____

4. _____

5. _____

6. _____

7. _____

8. _____

Exercise 13

FORMULA MASS, MOLES
AND MOLECULES

For the compound (or element) given, calculate the quantity indicated by the X. Determine the molecular or formula mass to the nearest hundreth of a gram per mole. For the other values use the appropriate number of significant figures.

See next page for exercise problems.

element or compound	formula mass (g/mole)	mass (grams)	moles	molecules or formula units	a	b	c	d	answer
1. sodium	X				45.98	22.99	11	none	_____
2.		11.5	X		0.500	1.05	1.00	none	_____
3.		11.5		X	8.31×10^{-25}	6.02×10^{23}	3.01×10^{23}	none	_____
4. fluorine (molecular)	X				19.00	9.50	9.00	none	_____
5.		76	X		2.0	1.0	8.44	none	_____
6.		76		X	6.0×10^{23}	1.2×10^{24}	3.0×10^{23}	none	_____
7. water	X				17.01	18.02	10.00	none	_____
8.			X	1.5×10^{22}	2.5×10^{-2}	4.0×10^{1}	0.45	none	_____
9.		X		1.5×10^{22}	0.45	7.2×10^{2}	1.4×10^{-3}	none	_____
10. ammonia	X				16.02	15.01	17.03	none	_____
11.		1.70		X	6.02×10^{24}	6.02×10^{22}	0.100	none	_____
12. nitrogen dioxide	X				46.01	44.01	78.01	none	_____
13.		X		3.0×10^{21}	0.23	9.2×10^{3}	5.0×10^{-3}	none	_____
14. calcium carbonate	100.09	X	5.0		2.0×10^{1}	5.0×10^{2}	5.0×10^{-2}	none	_____
15.			5.0	X	1.2×10^{23}	8.3×10^{-24}	3.0×10^{24}	none	_____

	element or compound	formula mass (g/mole)	mass (grams)	moles	molecules or formula units	a	b	c	d	answer
16.	potassium sulfide	X				71.16	110.26	103.22	none	_____
17.			5.0		X	2.7×10^{22}	1.3×10^{25}	7.5×10^{-26}	none	_____
18.	magnesium nitrate	X				86.31	148.32	210.33	none	_____
19.			X		1.8×10^{24}	0.11	3.0	0.33	none	_____
20.			X		1.8×10^{24}	4.4×10^{2}	29	2.0×10^{-2}	none	_____
21.	ammonium sulfate	X				114.10	132.14	210.16	none	_____
22.			3.0	X		3.9×10^{2}	44	2.3×10^{-2}	none	_____
23.			3.0	X	X	1.4×10^{22}	3.8×10^{-26}	1.8×10^{24}	none	_____
24.	glucose ($C_6H_{12}O_6$)	X				180.16	96.00	340.15	none	_____
25.			X	2.0×10^{-2}		1.1×10^{-4}	3.6	9.0×10^{3}	none	_____
26.				2.0×10^{-2}	X	1.2×10^{22}	3.0×10^{25}	3.3×10^{-26}	none	_____
27.	barium phosphate	X				232.31	369.65	601.96	none	_____
28.			8.0×10^{-2}		X	2.9×10^{25}	8.0×10^{19}	4.5×10^{27}	none	_____
29.	iron	55.85	X		1	9.274×10^{-23}	3.363×10^{25}	2.973×10^{-26}	none	_____
30.	helium	4.00	X		1	1.66×10^{-25}	6.64×10^{-24}	1.51×10^{23}	none	_____

Exercise 14

PERCENT COMPOSITION, EMPIRICAL AND MOLECULAR FORMULAS

Calculate the % composition by mass of the elements in the compounds given:

	compound	element	a	b	c	d	answer
1.	methanol (CH_4O)	%C =	41.39	37.49	23.08	none	_____
2.		%H =	12.58	15.38	3.47	none	_____
3.		%O =	61.54	55.14	16.67	none	_____
4.	aspirin ($C_9H_8O_4$)	%C =	60.00	42.85	57.44	none	_____
5.		%H =	38.10	4.48	8.51	none	_____
6.		%O =	35.52	34.04	19.05	none	_____
7.	calcium nitrate	%Ca =	39.26	26.71	24.42	none	_____
8.		%N =	17.07	13.71	9.33	none	_____
9.		%O =	47.02	58.50	63.96	none	_____
10.	ammonium sulfide	%N =	27.95	41.11	36.84	none	_____
11.		%H =	36.84	8.05	11.83	none	_____
12.		%S =	64.00	47.06	21.05	none	_____

Determine the empirical formula for the compounds with mass percents below:

element percents	a	b	c	d	answer
13. 79.88% Cu 20.12% O	Cu_2O	CuO	Cu_4O	none	_____
14. 68.42% Cr 31.58% O	Cr_2O_3	CrO_3	CrO_2	none	_____
15. 72.36% Fe 27.64% O	FeO	Fe_2O_3	Fe_3O_4	none	_____

Determine the molecular formulas from the following percent compositions by mass and molecular mass data.

element percents	mol. mass (g/mol)	a	b	c	d	answer
16. 92.26% C 7.74% H	78 ± 1	CH	$C_{12}H$	C_6H_6	none	_____
17. 40.00% C 6.71% H 53.29% O	60 ± 1	CH_2O	$C_2H_4O_2$	C_6HO_8	none	_____
18. 40.00% C 6.71% H 53.29% O	181 ± 2	$C_6H_{12}O_6$	CH_4O	$C_2H_4O_2$	none	_____
19. 9.93% C 58.64% Cl 31.43% F	121 ± 2	$CClF$	CCl_2F_2	CCl_6F_3	none	_____
20. 30.44% N 69.55% O	91 ± 2	NO	NO_2	N_2O_4	none	_____

Exercise 15

CHEMICAL REACTIONS: BALANCING AND CLASSIFICATION

Questions 1–48 on the following page refer to the unbalanced reactions A–P.

A.   __ $C_3H_8(g)$ +   __ $O_2(g)$ $\rightarrow$ __ $CO_2(g)$ +   __ $H_2O(g)$

B.   __ $C_2H_6O(l)$ +   __ $O_2(g)$ $\rightarrow$ __ $CO_2(g)$ +   __ $H_2O(g)$

C.   __ $N_2O_5(g)$ +   __ $H_2O(l)$ $\rightarrow$ __ $HNO_3(aq)$

D.   __ $N_2(g)$ +   __ $H_2(g)$ $\rightarrow$ __ $NH_3(g)$

E.   __ $Al(s)$ +   __ $HCl(aq)$ $\rightarrow$ __ $AlCl_3(aq)$ +   __ $H_2(g)$

F.   __ $CaCl_2(aq)$ +   __ $AgNO_3(aq)$ $\rightarrow$ __ $Ca(NO_3)_2(aq)$ +   __ $AgCl(s)$

G.   __ $K_2O(s)$ +   __ $H_2O(l)$ $\rightarrow$ __ $KOH(aq)$

H.   __ $Fe(s)$ +   __ $H_2O(l)$ $\rightarrow$ __ $Fe_3O_4(s)$ +   __ $H_2(g)$

I.   __ $Cl_2(aq)$ +   __ $NaI(aq)$ $\rightarrow$ __ $NaCl(aq)$ +   __ $I_2(aq)$

J.   __ $Fe(OH)_3(s)$ $\rightarrow$ __ $Fe_2O_3(s)$ +   __ $H_2O(g)$

K.   __ $BaCl_2(aq)$ +   __ $Na_3PO_4(aq)$ $\rightarrow$ __ $Ba_3(PO_4)_2(s)$ +   __ $NaCl(aq)$

L.   __ $C_6H_{12}O_6(s)$ +   __ $O_2(g)$ $\rightarrow$ __ $CO_2(g)$ __ $H_2O_{(g)}$

M.   __ $NaHCO_3(s)$ $\rightarrow$ __ $Na_2CO_3(s)$ +   __ $H_2O(g)$ +   __ $CO_2(g)$

N.   __ $Cu(s)$ +   __ $AgNO_3(aq)$ $\rightarrow$ __ $Cu(NO_3)_2(aq)$ +   __ $Ag(s)$

O.   __ $SO_2(g)$ +   __ $O_2(g)$ $\rightarrow$ __ $SO_3(g)$

P.   __ $KClO_3(s)$ $\rightarrow$ __ $KCl(s)$ +   __ $O_2(g)$

For questions 1-16, determine the lowest whole number sum of the coefficients. After balancing, $Na + H_2O \rightarrow NaOH + H_2$ becomes $2\,Na + 2\,H_2O = 2\,NaOH + H_2$ and the sum of the coefficients = 2 + 2 + 2 + 1 = 7.

For questions 17-32, classify reactions A-P as:

a. combination
b. decomposition
c. combustion
d. single replacement
e. double replacement

For questions 33-48, classify reactions A-P as:

a. redox
b. not redox

rxn.	#	a	b	c	d	answer	#	answer	#	answer
A.	1.	18	13	12	none	_____	17.	_____	33.	_____
B.	2.	8	9	12	none	_____	18.	_____	34.	_____
C.	3.	3	4	5	none	_____	19.	_____	35.	_____
D.	4.	6	7	5	none	_____	20.	_____	36.	_____
E.	5.	12	8	10	none	_____	21.	_____	37.	_____
F.	6.	4	5	6	none	_____	22.	_____	38.	_____
G.	7.	4	3	6	none	_____	23.	_____	39.	_____
H.	8.	11	12	16	none	_____	24.	_____	40	_____
I.	9.	4	6	5	none	_____	25.	_____	41.	_____
J.	10.	3	6	4	none	_____	26.	_____	42.	_____
K.	11.	9	10	12	none	_____	27.	_____	43.	_____
L.	12.	22	19	18	none	_____	28.	_____	44.	_____
M.	13.	6	4	5	none	_____	29.	_____	45.	_____
N.	14.	6	4	5	none	_____	30.	_____	46.	_____
O.	15.	3	5	4	none	_____	31.	_____	47.	_____
P.	16.	5	3	7	none	_____	32.	_____	48.	_____

Exercise 16

DOUBLE REPLACEMENT REACTIONS, NET IONIC EQUATIONS

Assume that you have aqueous solutions of each of the compounds below. Write balanced formula (FE), total ionic (TIE) and net ionic (NIE) equations for the reactions that would result from the mixing of each pair of solutions. In order to have a reaction, at least one of the products must be slightly ionized, insoluble (check solubility chart on page 399), or a gas or a compound that decomposes into a gas. If you do not think a reaction would occur, write "NAR" (no apparent reaction).

1. FE _____$NaIO_3(aq)$ + _____$SrCl_2(aq)$ =

   TIE

   NIE

2. FE _____$NaIO_3(aq)$ + _____$Ce(NO_3)_3(aq)$ =

   TIE

   NIE

3. FE _____$NaOH(aq)$ + _____$CoCl_2(aq)$ =

   TIE

   NIE

4. FE _____$NaOH(aq)$ + _____$HCl(aq)$ =

   TIE

   NIE

344

5. FE _____$Na_2SO_4(aq)$ + _____$CoCl_2(aq)$ =

 TIE

 NIE

6. FE _____$Na_3PO_4(aq)$ + _____$CoCl_2(aq)$ =

 TIE

 NIE

7. FE _____$Na_3PO_4(aq)$ + _____$HCl(aq)$ =

 TIE

 NIE

8. FE _____$Na_3PO_4(aq)$ + _____$SrCl_2(aq)$ =

 TIE

 NIE

9. calcium chloride + sodium carbonate =

 FE

 TIE

 NIE

10. copper(II) nitrate + sodium sulfate =

FE

TIE

NIE

11. ammonium chloride + potassium hydroxide =

FE

TIE

NIE

12. sodium bicarbonate + nitric acid =

FE

TIE

NIE

13. zinc chloride + sodium phosphate =

FE

TIE

NIE

14. lead(II) acetate + nitric acid =

FE

TIE

NIE

15. ammonium bromide + iron(III) nitrate =

FE

TIE

NIE

16. copper(II) sulfate + sodium hydroxide =

FE

TIE

NIE

17. silver nitrate + potassium chromate =

FE

TIE

NIE

Exercise 17

STOICHIOMETRY

For the reaction given, determine the quantity of the substance indicated by an X in the units indicated from the amount of the reactant used or product formed. When amounts of two reactants are given, it is possible that one is present in excess and the other is the limiting reagent. Be sure to take this into consideration.

A. moles to moles

					a	b	c	d	answer
$CH_4(g)$ +	$2 O_2(g)$	=	$CO_2(g)$ +	$2 H_2O(g)$					
1. 4.0 moles	X moles				4.0	8.0	2.0	none	_____
2. 4.0 moles	excess		X moles		4.0	8.0	2.0	none	_____
3. 4.0 moles	excess			X moles	4.0	8.0	2.0	none	_____
$C_3H_8(g)$ +	$5 O_2(g)$	=	$3 CO_2(g)$ +	$4 H_2O(g)$					
4.	X moles		0.60 moles		0.60	0.36	1.0	none	_____
5. X moles			0.60 moles		1.80	0.20	0.60	none	_____
6.			0.60 moles	X moles	0.60	0.80	0.45	none	_____
$4 FeS_2(s)$ +	$11 O_2(g)$	=	$2 Fe_2O_3(s)$ +	$8 SO_2(g)$					
7. 0.80 moles	X moles				0.80	0.29	2.2	none	_____
8. 0.80 moles	2.2 moles		X moles		0.40	1.6	0.80	none	_____
9. 0.80 moles	4.4 moles		X moles		0.40	1.6	0.80	none	_____

B. grams to grams

			a	b	c	d	answer

$Zn_{(s)}$ + $2 HCl_{(aq)}$ = $ZnCl_{2(aq)}$ + $H_{2(g)}$

			a	b	c	d	answer
10. 6.54 g	X grams		6.54	7.29	18.2	none	_____
11. 6.54 g	excess	X grams	13.6	6.54	27.2	none	_____
12. 6.54 g	excess	X grams	0.101	0.403	0.202	none	_____

$CaCl_{2(aq)}$ + $2 AgNO_{3(aq)}$ = $2 AgCl_{(s)}$ + $Ca(NO_3)_{2(aq)}$

			a	b	c	d	answer
13.	X grams	2.8 g	1.7	6.6	3.3	none	_____
14. X grams		2.8 g	1.1	2.2	4.3	none	_____

$3 Fe_{(s)}$ + $4 H_2O_{(l)}$ = $Fe_3O_{4(s)}$ + $4 H_{2(g)}$

			a	b	c	d	answer
15. 28 g	18 g	X grams	1.3	2.0	0.76	none	_____
16. 28 g	9.0 g	X grams	1.3	1.0	0.76	none	_____

GAS LAWS

For the problems on the next page, assume constant mass and calculate the missing quantity.

	V₁	P₁	T₁	V₂	P₂	T₂	a	b	c	d	answer
1.	1.5 L	1.00 atm	25°C	——	1.00 atm	100°C	6.0 L	1.9 L	1.2 L	none	____
2.	3.6 L	1.00 atm	——	2.1 L	1.00 atm	196 K	114 K	336 K	274 K	none	____
3.	0.45 L	1.00 atm	0°C	——	595 mm$_{Hg}$	0°C	0.35 L	2.7×10^2 L	0.57 L	none	____
4.	39 L	——	0°C	26 L	0.80 atm	0°C	1.2 atm	0.53 atm	0.80 atm	none	____
5.	22.4 L	760 mm$_{Hg}$	0°C	——	760 mm$_{Hg}$	25°C	20.5 L	22.4 L	24.5 L	none	____
6.	64.0mL	780 mm$_{Hg}$	37°C	——	520 mm$_{Hg}$	127°C	124 mL	74.4 mL	55.1 L	none	____
7.	7.3 L	——	23°C	1.9 L	750 mm$_{Hg}$	77 K	50.8 mm$_{Hg}$	1.11×10^5 mm$_{Hg}$	750 mm$_{Hg}$	none	____
8.	5.0 L	750 mm$_{Hg}$	20°C	6.0 L	720 mm$_{Hg}$	——	65 °C	338°C	93°C	none	____
9.	75 L	550 mm$_{Hg}$	25°C	55 L	——	50°C	437 mm$_{Hg}$	1.07 atm	692 mm$_{Hg}$	none	____
10.	——	1 atm	25°C	0.40 L	1.1 atm	0°C	0.33 L	0.40 L	0.48 L	none	____

11. How many moles of nitrogen gas are in a 2.00 liter vessel filled with nitrogen at 0.90 atm and 25°C?

a. 0.074 moles   b. 0.88 moles   c. 56 moles   d. none of the preceding

12. 7.20 grams of an unknown gas is found to have a volume of 4.00 L at 760 mm$_{Hg}$ and 25°C. What is the molecular mass of the gas?

a. 0.0579 g/mol   b. 44.0 g/mol   c. 3.69 g/mol   d. none of the preceding

Exercise 19

CHARLES' LAW (GRAPH)

In an earlier discussion of graphing techniques (pages 42-46), the volume of a gas at several temperatures was reported. A graph was included for the temperature range -40°C to 100°C and it was observed that the volume and temperature are linearly related or fit an equation $V = Mt + b$. The temperature at which the volume should go to zero is of considerable interest and value. To obtain that temperature, plot the volume on the vertical scale from 0 to 3 liters and the temperature on the horizontal scale from -300°C to 100°C. Draw the best straight line through the data and extend it down to the point where the volume is 0.00 liters.

temp. (°C)	volume (liters)
100	2.98
80	2.87
60	2.68
40	2.53
20	2.33
0	2.21
-20	2.07
-40	1.88

Questions

1. At what temperature does the volume go to zero (or what is absolute zero in degrees Celsius) according to your graph?

2. Explain any discrepancy between the value you obtained in *Question 1* above and the expected value of -273.15°C. Is it valid or accurate to extrapolate the data so far below the region of measurement?

3. Will the volume actually go to zero at absolute zero? Explain your answer.

4. According to your graph, what would the volume of the gas be at -200°C?

Exercise 20

VAPOR PRESSURE OF WATER VERSUS TEMPERATURE (GRAPH)

On the accompanying graph plot the vapor pressure of water on the vertical scale versus the temperature on the horizontal scale.

temp. (°C)	vapor pressure (mm Hg)	temp. (°C)	vapor pressure (mm Hg)
-10	2.1	80	355.1
0	4.6	85	433.6
10	9.2	90	525.8
20	17.5	95	633.9
30	31.8	98	707.3
40	55.3	100	760.0
50	92.5	102	815.9
60	149.4	105	906.1
70	233.7		

Questions

1. What is the vapor pressure of water at 25°C?

2. At what temperature should water boil in an open container if atmospheric pressure is 760 mm_{Hg}?

3. Atmospheric pressure at Denver is about 0.83 atm. What is the boiling point of water in Denver?

Exercise 21

CONCENTRATION CALCULATIONS AND SOLUTION STOICHIOMETRY

1. Determine the mass percent of glucose in a solution prepared by adding 5.0×10^1 g of glucose to 2.0×10^2 g of water.

   a. 25%   b. 20%   c. 500%   d. none of the answers given   _____

2. How much sodium hydroxide is needed to make 3.50×10^2 g of a 10% NaOH soln.

   a. 35 g   b. 32 g   c. 3.5×10^3 g   d. none   _____

3. What is the molarity of a solution prepared by dilution of 0.200 moles of potassium iodate to 5.00×10^2 mL with water?

   a. 4.00×10^{-4} M   b. 0.400 M   c. 0.100M   d. 2.50 M   e. none   _____

4. 6.0 g of acetic acid are diluted to 250 mL with water. What is the molarity of the solution?

   a. 0.40 M   b. 24 M   c. 4.0×10^{-3} M   d. 2.5 M   e. none   _____

5. How many grams of sodium hydroxide are needed to prepare 2.0×10^2 mL of a 0.30 M NaOH solution?

   a. 6.0×10^{-2} g   b. 60 g   c. 27 g   d. 2.4 g   e. none   _____

6. 25 mL of 6.0 M HCl are diluted to 5.0×10^2 mL with water. What is the molarity of the resulting solution?

   a. 0.30 M   b. 120 M   c. 3.3 M   d. none   _____

7. How many mL of 6.0 M NH_3 should be diluted to 250 mL with water to prepare a 0.050 M NH_3 solution?

   a. 3.0×10^4 mL   b. 2.1 mL   c. 1.2 mL   d. none   _____

8. 5.0 mL of 0.80 M silver nitrate solution are diluted to 1.0×10^2 mL with water. What is the molarity of the resulting solution?

   a. 1.6×10^{-2} M   b. 0.40 M   c. 4.0×10^{-2} M   d. none   _____

9. How many grams of potassium iodide are needed to prepare 1.0×10^2 mL of a 0.20 M potassium iodide solution?

   a. 0.33 g   b. 3.3×10^2 g   c. 83 g   d. 2.0×10^{-2} g   e. none   _____

10. 1.0×10^2 mL of a saturated aqueous solution of iodine at 25°C contains 0.030 g of iodine. What is the molarity of a saturated solution?

   a. 0.30 M   b. 2.4×10^{-3} M   c. 1.2×10^{-3} M   d. none   _____

11. How many mL of 0.50 M HCl should be diluted to 5.0×10^2 mL with water to prepare a 0.025 M HCl solution?

   a. 25 mL   b. 2.5 ml   c. 1.0×10^1 mL   d. none   _____

12. 16.25 mL of 0.1012 M NaOH are needed to titrate 20.00 mL of an HCl solution to the end point. What is the molarity of the hydrochloric acid solution?

   a. 0.1246 M   b. 0.08222 M   c. 8.029 M   d. none   _____

13. 22.47 mL of 0.2104 M NaOH are needed to titrate 15.00 mL of an H_2SO_4 solution to the end point. What is the molarity of the sulfuric acid solution?

   a. 0.1576 M   b. 0.6304 M   c. 0.3152 M   d. none   _____

14. 11.27 mL of 0.1523 M NaOH are needed to titrate 0.350 g of a monoprotic acid to its end point. What is the molecular mass of the acid?

   a. 38.0 g/mol   b. 204 g/mol   c. 2.04×10^5 g/mol   d. none   _____

15. 17.73 mL of an NaOH solution are needed to titrate 0.4532 g of a monoprotic acid (molecular mass = 122.1 g/mol) to the end point. What is the molarity of the NaOH solution?

   a. 0.3204 M   b. 0.08035 M   c. 0.2093 M   d. none   _____

Exercise 22

ACIDS AND BASES, pH

For the reactions given, pick the letter of the compound that best fits the definition of Bronsted acids (proton donor) and Bronsted bases (proton acceptor) and their conjugates.

$$HNO_3(aq) \quad + \quad H_2O(l) \quad = \quad H_3O^+ \quad + \quad NO_3^-$$
$$\;\;a \qquad\qquad\quad b \qquad\qquad\quad c \qquad\qquad\; d$$

1. Bronsted acid _____

2. Bronsted base _____

3. Conjugate acid _____

4. Conjugate base _____

$$NH_3(aq) \quad + \quad H_2O(l) \quad = \quad NH_4^+ \quad + \quad OH^-$$
$$\;a \qquad\qquad\quad b \qquad\qquad\quad c \qquad\qquad\; d$$

5. Bronsted acid _____

6. Bronsted base _____

7. Conjugate acid _____

8. Conjugate base _____

$$CO_3^{2-} \quad + \quad H_2O(l) \quad = \quad HCO_3^- \quad + \quad OH^-$$
$$\;a \qquad\qquad\quad b \qquad\qquad\quad c \qquad\qquad\; d$$

9. Bronsted acid _____

10. Bronsted base _____

11. Conjugate acid _____

12. Conjugate base _____

357

$$\underset{a}{H_2PO_4^-} \quad + \quad \underset{b}{H_2O_{(l)}} \quad = \quad \underset{c}{H_3O^+} \quad + \quad \underset{d}{HPO_4^{2-}}$$

13. Bronsted acid _____

14. Bronsted base _____

15. Conjugate acid _____

16. Conjugate base _____

$$\underset{a}{HC_2H_3O_{2(aq)}} \quad + \quad \underset{b}{PO_4^{3-}} \quad = \quad \underset{c}{HPO_4^{2-}} \quad + \quad \underset{d}{C_2H_3O_3^-}$$

17. Bronsted acid _____

18. Bronsted base _____

19. Conjugate acid _____

20. Conjugate base _____

Calculate the quantity indicated by an X.

	[H⁺] (moles/L)	pH	pOH	[OH⁻] (moles/L)	a	b	c	d	answer
21.	1×10^{-3}	X			-3	3	11	none	_____
22.	1×10^{-3}		X		-3	3	11	none	_____
23.	1×10^{-3}			X	1×10^{-3}	11	1×10^{-11}	none	_____
24.	X	5			-5	1×10^{-5}	1×10^{-9}	none	_____
25.		5	X		-5	9	1×10^{-9}	none	_____
26.		5		X	1×10^{-5}	9	1×10^{-9}	none	_____
27.	X		11		1×10^{-3}	1×10^{-11}	3	none	_____
28.		X	11		1×10^{-3}	3	11	none	_____
29.			11	X	-3	1×10^{-11}	1×10^{-3}	none	_____
30.	X			1×10^{-4}	1×10^{-10}	1×10^{-4}	10	none	_____
31.		X		1×10^{-4}	4	1×10^{-4}	1×10^{-10}	none	_____
32.			X	1×10^{-4}	4	10	1×10^{-10}	none	_____
33.	X	9			1×10^{-5}	-9	1×10^{-9}	none	_____
34.		9	X		1×10^{-5}	5	-5	none	_____
35.		9		X	1×10^{-5}	1×10^{-9}	5	none	_____
36.	1×10^{-1}	X			-1	13	1	none	_____
37.	1×10^{-1}		X		1	13	1×10^{-13}	none	_____
38.	1×10^{-1}			X	-13	13	1×10^{-13}	none	_____
39.	X			1×10^{-2}	1×10^{-12}	-12	1×10^{-2}	none	_____
40.		X		1×10^{-2}	12	2	-12	none	_____
41.			X	1×10^{-2}	12	2	-12	none	_____

Exercise 23

OXIDATION STATES OF THE ELEMENTS AND NOMENCLATURE

A.   Give the oxidation states for the elements listed when they are present as ions in the chloride or sodium compounds given. Remember that compounds are electrically neutral so the oxidation states of the ions in a compound must sum to zero.   Also chloride in binary compounds always has an oxidation number of -1 (e.g., for $SnCl_4$, tin must be +4 to neutralize the 4 negative chloride ions: $x + 4(-1) = 0, x = 4$) and sodium ion is always +1. Also fill in the blanks for systematic and common names of the ions when indicated (Sn^{4+} = tin(IV) = stannic).   Please recognize that this list includes only the most commonly encountered ions of the elements given and in some cases other oxidations states are found.

multiple choice answers: a. 1   b. 2   c. 3   d. 4   e. 5

CHLORIDES

#	element symbol	atomic #	formula of chloride	systematic name	common name	multiple choice answer
1.	H	1	HCl	_____		_____
2.	Li	3	LiCl	_____		_____
3.	Be	4	$BeCl_2$	_____		_____
4.	Na	11	NaCl	_____		_____
5.	Mg	12	$MgCl_2$	_____		_____
6.	Al	13	$AlCl_3$	_____		_____
7.	K	19	KCl	_____		_____
8.	Ca	20	$CaCl_2$	_____		_____
9.	Cr	24	$CrCl_3$	_____	_____	_____
10.	Mn	25	$MnCl_2$	_____	_____	_____

362

multiple choice answers: a. 1   b. 2   c. 3   d. 4   e. 5

CHLORIDES

#	element symbol	atomic #	formula of chloride	systematic name	common name	multiple choice answer
11.	Fe	26	$FeCl_2$	_____	_____	_____
12.	Fe	26	$FeCl_3$	_____	_____	_____
13.	Co	27	$CoCl_2$	_____	_____	_____
14.	Ni	28	$NiCl_2$	_____	_____	_____
15.	Cu	29	$CuCl$	_____	_____	_____
16.	Cu	29	$CuCl_2$	_____	_____	_____
17.	Zn	30	$ZnCl_2$	_____		_____
18.	Rb	37	$RbCl$	_____		_____
19.	Sr	38	$SrCl_2$	_____		_____
20.	Ag	47	$AgCl$	_____		_____
21.	Sn	50	$SnCl_2$	_____	_____	_____
22.	Sn	50	$SnCl_4$	_____	_____	_____
23.	Cs	55	$CsCl$	_____		_____
24.	Ba	56	$BaCl_2$	_____		_____
25.	Ce	58	$CeCl_3$	_____	_____	_____
26.	Hg	80	Hg_2Cl_2	_____	_____	_____
27.	Hg	80	$HgCl_2$	_____	_____	_____
28.	Pb	82	$PbCl_2$	_____	_____	_____

multiple choice answers: a. 0   b. -1   c. -2   d. -3   e. none

SODIUM COMPOUNDS

#	element symbol	atomic #	formula of Na cmpd.	systematic name	multiple choice answer
29.	H	1	NaH	_____	_____
30.	O	8	Na_2O	_____	_____
31.	F	9	NaF	_____	_____
32.	S	16	Na_2S	_____	_____
33.	Cl	17	NaCl	_____	_____
34.	Br	35	NaBr	_____	_____
35.	I	53	NaI	_____	_____

Based on the answers in numbers 1-35 above, with the use of a periodic table, answer the following questions.

36. The oxidation state of Group IA (or 1) elements in compounds is always (except for H):

   a. 0   b. 1   c. 2   d. 3   e. none   _____

37. The oxidation state of Group IIA (or 2) elements in compounds is always:

   a. 0   b. 1   c. 2   d. 3   e. none   _____

38. The oxidation state of Group VIIA (or 17) elements in binary compounds is always (except dihalogen compounds - e.g. ICl):

   a. 0   b. 1   c. -1   d. -2   e. none   _____

39. The oxidation state of the transition metals in the problems above is (are):

   a. 0   b. 1   c. 2   d. 3   e. sometimes 1 but usually 2 or 3   _____

40. Write the electron configurations of the ions of IA, IIA and VIIA (or 1, 2 and 17) Group elements and compare them to the electron configurations of the inert gases. What generalization can be made for these ions?

B. Determine the oxidation number of the element indicated. Refer to your textbook for rules regarding the determination of oxidation numbers.

element	a	b	c	d	answer
1. iron in Fe_2O_3	3	2	4	none	_____
2. antimony in Sb_2O_5	2	-3	5	none	_____
3. bismuth in $BiCl_3$	5	-3	3	none	_____
4. nitrogen in NO_3^-	-3	5	6	none	_____
5. nitrogen in NO_2	-3	2	4	none	_____
6. nitrogen in NO_2^-	-3	3	4	none	_____
7. nitrogen in NO	0	-3	1	none	_____
8. nitrogen in N_2O	0	1	2	none	_____
9. nitrogen in N_2	0	-3	2	none	_____
10. nitrogen in NH_3	3	0	-3	none	_____
11. chlorine in ClO_4^-	-1	8	7	none	_____
12. chlorine in ClO_3^-	6	5	-1	none	_____
13. chlorine in ClO_2^-	3	4	-1	none	_____
14. chlorine in ClO^-	1	2	-1	none	_____
15. chlorine in Cl_2	-1	0	1	none	_____
16. chlorine in $NaCl$	-1	0	7	none	_____
17. uranium in $U(SO_4)_2$	0	2	4	none	_____
18. chromium in $Cr_2O_7^{2-}$	6	-6	7	none	_____
19. chromium in CrO_4^{2-}	6	-6	7	none	_____
20. iodine in KIO_3	6	-1	5	none	_____
21. phosphorous in Na_3PO_4	6	5	4	none	_____

Exercise 24

OXIDATION-REDUCTION
HALF REACTIONS

Write balanced half reactions and complete reactions for the following oxidation-reduction reactions.

1. copper(s) + silver ion(aq) → copper(II) ion(aq) + silver(s)

    oxid. half rxn.:

    red. half rxn.:

    overall rxn.:

2. iron(III)(aq) + iodide(aq) → iron(II)(aq) + iodine(s)

    oxid. half rxn.:

    red. half rxn.:

    overall rxn.:

3. iodide(aq) + persulfate(aq) $(S_2O_8^{2-})$ → iodine(aq) + sulfate(aq)

    oxid. half rxn.:

    red. half rxn.:

    overall rxn.:

4. aluminum(s) + hydrogen ion(aq) → aluminum ion(aq) + hydrogen(g)

    oxid. half rxn.:

    red. half rxn.:

    overall rxn.:

5. sodium(s) + water → sodium hydroxide(aq) + hydrogen(g)

   oxid. half rxn.:

   red. half rxn.:

   overall rxn.:

6. magnesium(s) + zinc ion(aq) → magnesium ion(aq) + zinc(s)

   oxid. half rxn.:

   red. half rxn.:

   overall rxn.:

The following oxidation-reduction reactions require the addition of H_2O and H^+ to balance some of the half reactions.

7. chloride + permanganate → manganese(II) ion + chlorine(g)   (aqueous acid)

   oxid. half rxn.:

   red. half rxn.:

   overall rxn.:

8. tin(II) ion + dichromate → tin(IV) ion + chromium(III) ion   (aqueous acid)

   oxid. half rxn.:

   red. half rxn.:

   overall rxn.:

9. iodide + iodate → iodine(aq)   (aqueous acid)

   oxid. half rxn.:

   red. half rxn.:

   overall rxn.:

Exercise 25

EQUILIBRIUM CONSTANTS, LE CHATELIER'S PRINCIPLE

Choose the correct equilibrium expression for the following reactions:

1. $2NO_2(g) = N_2O_4(g)$ K = _____

    a. $[NO_2]^2$ b. $\dfrac{[N_2O_4]}{[NO_2]^2}$ c. $\dfrac{[NO_2]^2}{[N_2O_4]}$ d. none

2. $N_2(g) + O_2(g) = 2\,NO(g)$ K = _____

    a. $\dfrac{[NO]^2}{[N_2][O_2]}$ b. $\dfrac{[NO]}{[N_2][O_2]}$ c. $\dfrac{[N_2][O_2]}{[NO]^2}$ d. none

3. $HF(aq) = H^+ + F^-$ K_a = _____

    a. $\dfrac{[HF]}{[H^+][F^-]}$ b. $[HF]$ c. $\dfrac{[H^+][F^-]}{[HF]}$ d. none

4. $CN^- + H_2O(l) = HCN(aq) + OH^-$ K_{hyd} = _____

    a. $\dfrac{[HCN][OH^-]}{[CN^-]}$ b. $[CN^-][H_2O]$ c. $\dfrac{[CN^-]}{[HCN][OH^-]}$ d. none

5. $HC_2H_3O_2(aq) + C_2H_5OH(l) = C_4H_8O_2(l) + H_2O(l)$ K = _____
   acetic acid   ethanol   ethyl acetate

    a. $\dfrac{[HC_2H_3O_2][C_2H_5OH]}{[C_4H_8O_2][H_2O]}$ b. $\dfrac{[C_4H_8O_2][H_2O]}{[HC_2H_3O_2][C_2H_5OH]}$

    c. $[HC_2H_3O_2][C_2H_5OH]$ d. none

6. $CaCO_3(s) = Ca^{2+} + CO_3^{2-}$ K_{sp} = _____

    a. $\dfrac{[Ca^{2+}][CO_3^{2-}]}{[CaCO_3]}$ b. $[Ca^{2+}][CO_3^{2-}]$ c. $[CaCO_3]$ d. none

The answers for the questions below are: a. left   b. right   c. no change

Assume an aqueous HF solution (#3 above) is at equilibrium. In which direction will the system shift to reachieve equilibrium if:

7. HCl is added to the system?   _____

8. NaF is added to the system?   _____

9. NaOH is added to the system?   _____

Assume the acetic acid, ethanol, ethyl acetate, water system (#5) is at equilibrium. In which direction will the system shift to reachieve equilibrium if:

10. acetic acid is added to the system?   _____

11. ethanol is added to the system?   _____

12. ethyl acetate is removed from the system?   _____

Assume that you have a saturated calcium carbonate solution (#6). In which direction will the system shift to reachieve equilbrium if:

13. sodium carbonate is added to the system?   _____

14. calcium carbonate is added to the system?   _____

15. HCl is added to the system?   _____

Exercise 26

NUCLEAR REACTIONS

Balance the following nuclear reactions by filling in the boxes. (Assume each of these reactions yields only two products.) The time periods by the arrows are the decay half-lives.

						$\underline{a}$	$\underline{b}$	$\underline{c}$	$\underline{d}$	$\underline{answer}$

1. $^{238}_{92}U$ $\xrightarrow{4.5\times10^{9}\ years}$ $^{234}_{90}Th$ + $\boxed{}$ $^{1}_{1}p$ $^{4}_{2}He$ $^{1}_{0}n$ none _____

2. $^{0}_{-1}e$ + $\boxed{}$ $\xleftarrow{24\ days}$ $^{234}_{89}Ac$ $^{234}_{91}Th$ $^{234}_{91}Pa$ none _____

3. $\xrightarrow{1.2\ min.}$ $^{234}_{92}U$ + $\boxed{}$ $^{0}_{-1}e$ $^{0}_{1}e$ $^{1}_{1}p$ none _____

4. $^{4}_{2}He$ + $\boxed{}$ $\xleftarrow{2.5\times10^{5}\ years}$ $^{238}_{94}Pu$ $^{230}_{92}U$ $^{230}_{90}Th$ none _____

5. $\xrightarrow{8.0\times10^{4}\ years}$ $^{226}_{88}Ra$ + $\boxed{}$ $^{4}_{2}He$ $^{2}_{2}He$ $^{0}_{-1}e$ none _____

6. $\boxed{}$ + $^{222}_{86}Rn$ $\xleftarrow{1.6\times10^{3}\ years}$ $^{4}_{2}He$ $^{0}_{-1}e$ $^{1}_{1}p$ none _____

7. $\xrightarrow{3.8\ days}$ $\boxed{}$ + $^{4}_{2}He$ $^{218}_{84}Rn$ $^{218}_{84}Po$ $^{218}_{88}Ra$ none _____

8. $\boxed{}$ + $^{214}_{82}Pb$ $\xleftarrow{30\ min.}$ $^{0}_{-1}e$ $^{4}_{4}Be$ $^{2}_{2}He$ none _____

9. $\xrightarrow{27\ min.}$ $\boxed{}$ + $^{0}_{-1}e$ $^{214}_{81}Tl$ $^{214}_{83}Bi$ $^{215}_{82}Pb$ none _____

10. $\boxed{}$ + $^{214}_{84}Po$ $\xleftarrow{20\ min.}$ $^{1}_{1}p$ $^{0}_{-1}e$ $^{1}_{0}n$ none _____

11. $\xrightarrow{1.6\times10^{-4}\ sec.}$ $\boxed{}$ + $^{4}_{2}He$ $^{210}_{82}Pb$ $^{210}_{82}Po$ $^{218}_{86}Rn$ none _____

12. $\boxed{}$ + $^{210}_{83}Bi$ $\xleftarrow{21\ years}$ $^{1}_{1}p$ $^{0}_{-1}e$ $^{0}_{1}e$ none _____

13. $\xrightarrow{5\ days}$ $\boxed{}$ + $^{0}_{-1}e$ $^{210}_{84}Po$ $^{209}_{83}Bi$ $^{210}_{82}Pb$ none _____

14. $\boxed{}$ + $^{206}_{82}Pb$ $\xleftarrow{138\ days}$ $^{4}_{2}He$ $^{2}_{2}He$ $^{4}_{4}Be$ none _____

stable

369

	a	b	c	d	answer

15. $^{7}_{3}Li$ + $^{1}_{1}H$ $\rightarrow$ ☐ + $^{1}_{0}n$ $^{7}_{4}Be$ $^{6}_{2}He$ $^{8}_{4}Be$ none   _____

16. $^{31}_{15}P$ + $^{2}_{1}H$ $\rightarrow$ $^{32}_{15}P$ + ☐ $^{4}_{2}He$ $^{1}_{1}H$ $^{3}_{1}H$ none   _____

17. $^{239}_{94}Pu$ + $^{2}_{1}H$ $\rightarrow$ ☐ + $^{1}_{0}n$ $^{240}_{93}Np$ $^{241}_{94}Pu$ $^{240}_{95}Am$ none   _____

18. $^{239}_{94}Pu$ + $^{4}_{2}He$ $\rightarrow$ ☐ + $^{1}_{0}n$ $^{242}_{96}Cm$ $^{244}_{95}Am$ $^{243}_{96}Cm$ none   _____

19. How many years would it take for $^{238}_{92}U$ to decay from a total of 64 grams to 8 grams?

    a. 4.5×10^{9} years   b. 1.35×10^{10} years   c. 9.0×10^{9} years

    d. 1.80×10^{10} years   e. none of the previous answers   _____

20. Starting with 800 grams of $^{214}_{82}Pb$, how much would you have left after 135 minutes?

    a. 25 g   b. 50 g   c. 400 g   d. 12.5 g   e. none of the previous answers   _____

Exercise 27

ORGANIC FUNCTIONAL GROUPS, NOMENCLATURE OF SATURATED HYDROCARBONS

Classify each of the following compounds by the functional group present (e.g., CH_3CH_2OH - alcohol). Choose the correct multiple choice answer. Mark both code letters when appropriate.

a.	alkane	e.	alcohol	ae.	ketone
b.	alkene	ab.	ether	bc.	carboxylic acid
c.	alkyne	ac.	amine	bd.	ester
d.	aromatic	ad.	aldehyde	be.	amide

1. $CH_3CH_2CH_2NH_2$ _____

2. $CH_3CH_2\overset{\displaystyle O}{\overset{\|}{C}}CH_3$ _____

3. $CH_3CH_2CH_3$ _____

4. $CH_3CH_2\overset{\displaystyle O}{\overset{\|}{C}}OH$ _____

5. $CH_3CH{=}CH_2$ _____

6. $CH_3\overset{\displaystyle O}{\overset{\|}{C}}OCH_3$ _____

7. $CH_3CH_2\overset{\displaystyle O}{\overset{\|}{C}}H$ _____

8. $CH_3C{\equiv}CH$ _____

9. 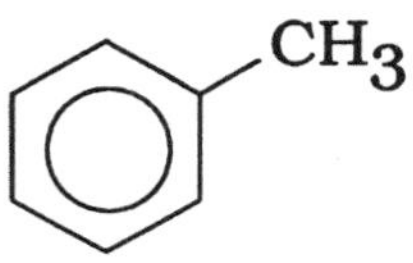
_____

10. $CH_3CH_2\overset{\displaystyle OH}{\overset{\|}{C}H}CH_3$ _____

11. $CH_3CH_2\overset{\displaystyle O}{\overset{\|}{C}}NH_2$ _____

12. $CH_3CH_2OCH_2CH_3$ _____

From the choices below select the best name for the structures of the saturated hydrocarbons given.

a.   4-*t*-butyloctane
b.   1,1-dimethylbutane
c.   2,3-dimethylhexane
d.   4,5-dimethylhexane
e.   2,2-dimethyl-3-propylheptane
ab. 3-ethylhexane
ac. 3-ethyl-2-methyl-4-propylheptane
ad. 4-*iso*propylheptane

ae. 2-methylbutane
bc. 3-methylbutane
bd. 2-methylpentane
be. 2-methyl-3-propylhexane
cd. 3-propylpentane
ce. 2-propylpropane
de. 2,2,4-trimethylpentane
abc. 2,4,4-trimethylpentane

13. 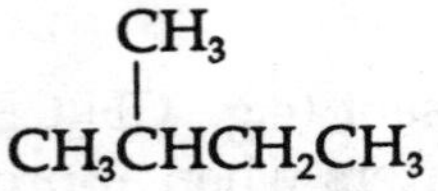
CH₃
|
CH₃CHCH₂CH₃   _____

14. 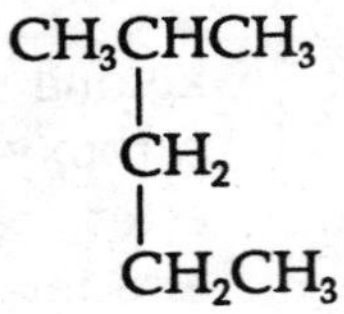
CH₃CHCH₃
|
CH₂
|
CH₂CH₃   _____

15. 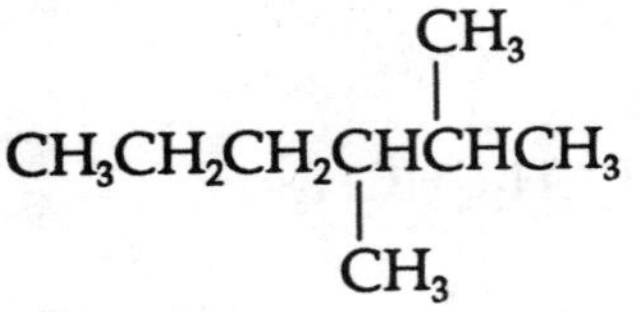
           CH₃
           |
CH₃CH₂CH₂CHCHCH₃   _____
              |
              CH₃

16. 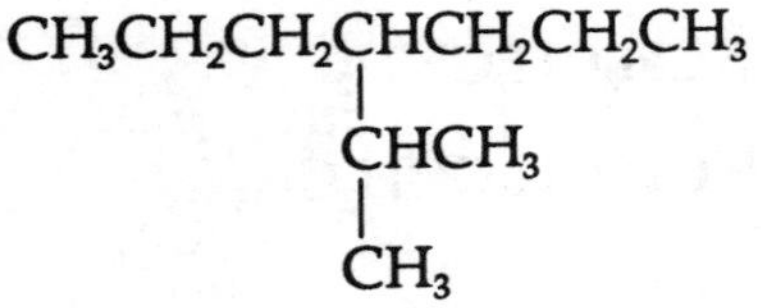
CH₃CH₂CH₂CHCH₂CH₂CH₃ _____
         |
         CHCH₃
         |
         CH₃

17. 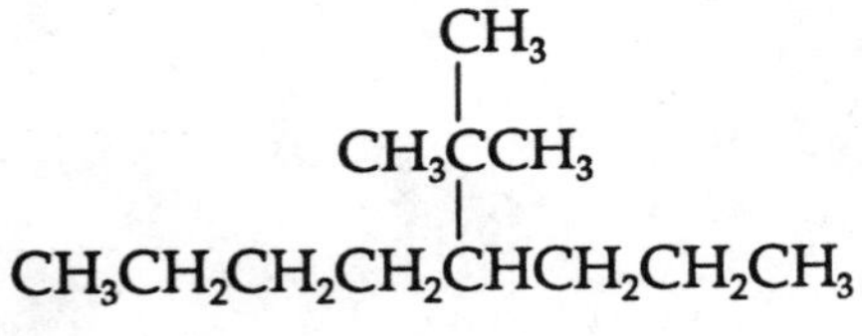
            CH₃
            |
            CH₃CCH₃
            |
CH₃CH₂CH₂CH₂CHCH₂CH₂CH₃ _____

18. 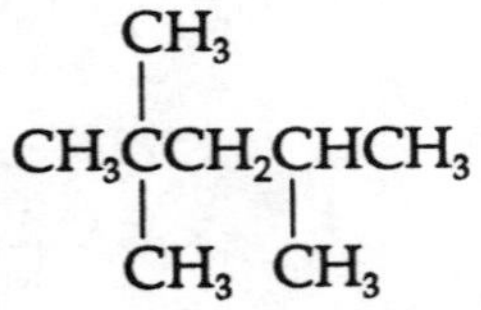
   CH₃
   |
CH₃CCH₂CHCH₃   _____
   | |
   CH₃   CH₃

19. 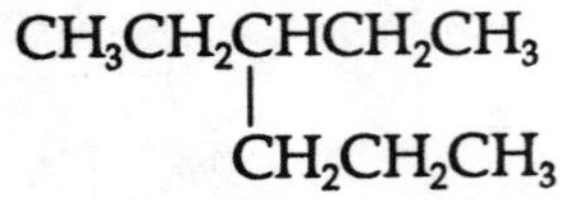
CH₃CH₂CHCH₂CH₃   _____
        |
        CH₂CH₂CH₃

20. 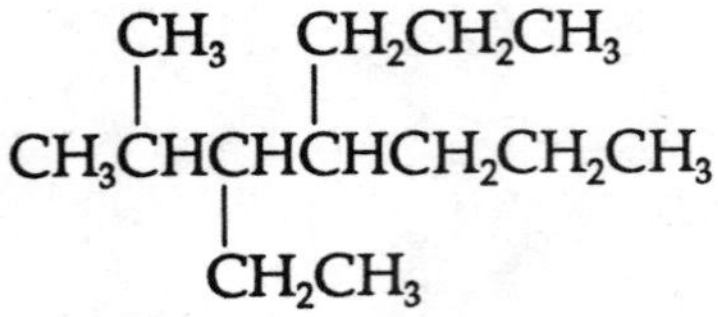
   CH₃   CH₂CH₂CH₃
   | |
CH₃CHCHCHCH₂CH₂CH₃   _____
      |
      CH₂CH₃

Exercise 28

LABORATORY TERMINOLOGY

Match the word or code of the words below with the definitions that follow (or darken the correct code on a Scantron form):

a	accuracy	be	extraction	bcd	saturated
b	anhydrous	cd	filtration	bce	solute
c	aqueous	ce	heterogeneous	bde	solution
d	chromatography	de	homogeneous	cde	solvent
e	condensate	abc	hygroscopic	abcd	spectroscopy
ab	conductivity	abd	immiscible	abce	sublimation
ac	deliquescence	abe	miscible	abde	supersaturated
ad	desiccant	acd	precipitate	acde	synthesis
ae	distillation	ace	precision	bcde	titration
bc	efflorescence	ade	recrystallization	abcde	unsaturated
bd	evaporation				

1. process of losing waters of hydration _____

2. insoluble solid that forms as a result of the mixing of two solutions or as a result of a solvent or temperature change _____

3. change of state directly from solid to gas _____

4. indicates that the solvent is water _____

5. having the same physical composition and properties throughout _____

6. one or more substances dissolved in a host medium _____

7. the preparation of a substance from one or more other substances using chemical reactions in the laboratory _____

8. a measure of the ability of a substance to carry electricity or transfer heat _____

9. used to describe two liquids that are insoluble in each other and form two phases or layers when mixing is attempted _____

10. drawing of a solute into a solvent in which it is more soluble than it is in the original solvent _____

11. process of a compound absorbing water vapor from the air _____

12. the liquid that results when the vapor of a substance is cooled below its boiling point _____

13. degree of conformity of a measurement to a standard value _____

14. an unstable condition in which the solvent contains more dissolved
     solute than can remain dissolved under normal conditions _____

15. used to describe two liquids that are soluble in each other and
     therefore form one phase when mixed _____

16. a mixture containing discernible particles having different composition
     or physical properties from some other portion of the mixture _____

17. a substance that dissolves a solute to form a homogeneous mixture _____

18. the agreement between several measurements of the same quantity
     or the number of significant figures in a measurement _____

19. observation of the wavelength and intensity of light absorbed or
     emitted by a substance _____

20. describes a solution in which as much of the solute is dissolved
     as possible under prevailing conditions _____

21. a separation process in which a liquid is converted to the vapor
     phase and then condensed back into the liquid phase _____

22. literally means without water and generally used to describe samples
     of inorganic compounds that do not contain waters of hydration _____

23. a method for determining volumetrically the concentration of a substance
     in solution using another substance of known concentration _____

24. a compound used to minimize the amount of water vapor in a container
     by virtue of its ability to absorb water vapor from the air _____

25. used to describe a solution in which more of the solute can be
     dissolved without heating _____

26. used to describe substances that have a strong tendency to absorb
     water vapor from the air _____

27. the process of separating solids from a liquid by passing the mixture
     through a porous barrier _____

28. the change of a substance from the liquid or solid phase the to
     vapor phase _____

29. a homogeneous mixture _____

30. a purification method used for solids that involves dissolving of a
     substance in a hot solvent, cooling of the solution and collection
     of the resulting crystals by filtration _____

31. a separation process in which the components of a mixture separate
     because of their different preferences for a moving phase and a
     stationary phase _____

LABORATORY TECHNIQUE REVIEW

For simplification and ease of learning and study, chemistry is classified into narrower fields. Probably the most common classification system includes organic, inorganic, analytical and physical chemistry branches. Another classifcation system focuses more on the goals and operations: synthesis, separation and purification, analysis or identification and applications. For the purposes of this review, the questions will be divided according to the latter classification system.

A. Synthesis

1. Suggest reactions and techniques that could be used to prepare each of the following gases:

    a.   hydrogen

    b.   oxygen

    c.   carbon dioxide

2. Suggest double replacement reactions and techniques that could be used to prepare the following compounds:

    a.   lead sulfate

    b.   strontium oxalate

3. 15 g of lithium aluminum hydride reacts with 35 g of water producing 3.00 g of hydrogen. Balance the equation below and calculate the theoretical and percent yields of hydrogen gas.

    ____$LiAlH_4$(s) + ____H_2O(s) = ____$LiOH$(s) + ___$Al(OH)_3$(s) + ___H_2(g)

B. Separation and purification.

1. During this course, in attempting to separate and purify various samples, you have used the techniques of evaporation, filtration (gravity and vacuum), recrystallization, extraction, paper chromatography and distillation. For each problem below, state which method you would try first. Explain your choice.

   a. Recovery of both salt and water from saturated salt water

   b. 2 grams of aspirin contaminated with 5% of unknown organic compounds

   c. After a small amount of protein (0.01 grams) is isolated from a milk sample, the amide linkages are broken to liberate the component amino acids

   d. Isolation of water and acetone from 10 mL of a 1:1 mixture

   e. Isolation of sand and salt from 5 grams of a 1:1 mixture

2. Addition of a saturated lithium hydroxide solution to a 33.28 g evaporating dish results in a total mass of 47.53 g. After evaporation to dryness, the dish and contents weigh 34.69 g. Determine the mass percent of lithium hydroxide in the saturated solution. Also by comparison to the solubility of lithium hydroxide reported in *Appendix C*, calculate the percentage deviation of your calculated value.

C. **Analysis or identification.**

1. Intensive physical properties and chemical properties ane useful when attempting to verify the identity of a compound or determine the structure of a new compound or determine the amounts of compounds in a mixture. During this course, some of the properties you have used are solubility, density, melting point, boiling point, color, flame test results and quantitative techniques including titration, pH and spectroscopic measurements. For each problem below, state which technique you would use. Explain your choice.

   a. Distinguish between aqueous solutions of copper sulfate, cobalt nitrate and calcium chloride

   b. Determine the concentration of copper(II) in a solution three ways

   c. Distinguish between ethanol and water (2 ways)

   d. Distinguish between 0.1000 M HCl and 0.1100 M HCl

   e. Distinguish between 0.1 M HCl and 0.1 M $HC_2H_3O_2$

   f. Distinguish between gold and fools gold (iron pyrite - FeS_2)

   g. Distinguish between diamonds and zircons

h. Distinguish between aspirin (acetylsalicylic acid) and acetominophen

i. Distinguish between solutions containing sodium and potassium ions

j. Distinguish between solutions containing silver and zinc ions

2. When a 52.75 g cylinder of an unknown lustrous element is immersed in 22.5 mL of water. the water level rises to 27.5 mL. The cylinder has a length of 4.25 cm and a diameter of 1.226 cm.

   a. Calculate the density of the element two different ways.

   b. A small piece of the element does not react with a copper sulfate solution. Discuss the significance of this result.

   c. What is the identity of the element?   Explain how you came to your conclusion.

3. It takes 11.31 mL of 0.2111 M NaOH to titrate 0.325 g of a monoprotic acid to the end point. What is the formula mass of the acid?   Could the acid be one of the following: $NaHSO_4$, $KHSO_4$, $HC_2H_3O_2$? Explain your answer.

4. 11.18 mL of 0.1050 M sodium thiosulfate are required to titrate 0.273 g of a copper compound to the end point. Calculate the mass percent of copper in the sample. Could the compound be copper nitrate hemipentahydrate? Explain your answer.

D. Applications.

1. Material Selection. In *Exercise 11*, one of the questions dealt with the selection of materials for specific applications. In that case however, the instructions limited you to elements. This time, without any restrictions, state the criteria you would use to select a material and suggest possible materials.

    a. antifreeze solution for cars

    b. bicycle frames

    c. food storage bags

    d. a coating for magnetic spin bars

    e. soda can material

    f. food additives for processed meat

2. Societal Issues. Each of the following is a controversial issue currently confronting society. Give the arguments pro and con on each side of the issue and state your preference.

    a. Should metropolitan water supplies be fluoridated?

    b. Should more nuclear fission power plants be constructed?

    c. Should we continue to expand the world's consumption of fossil fuels or should we begin a phase out of fossil fuels?

    d. Should we invest heavily in nuclear fusion research?

    e. Should we continue to do research in biotechnology?

    f. Should people take vitamin C on a daily basis?

    g. Should the ban on freons be reversed?

    h. Should a college cafeteria use paper plates, styrofoam plates or reuseable dishware?

i. Should nitrites be added to bacon, hot dogs, bologna, etc.?

j. Should dentists use silver amalgams to fill cavities?

k. Should insecticides and herbicides be used in agriculture?

E. Safety and an Extension.

1. Reread *page 4* of this book and answer the question on that page again.

2. What was the "greenish-blue liquid" that formed? (Hint: It may be necessary to refer to a general chemistry book to obtain answers to questions 2, 3 and 4.)

3. Why did the air become "dark red" and is the gas that caused this color very toxic?

4. Write balanced formula and net ionic equations for the reaction of copper with nitric acid.

5. Assume that a copper penny has a mass of 3.1 grams and calculate the volume in milliliters of concentrated nitric acid (16 M) needed to react stoichiometrically with the copper. Also calculate the number of grams and the volume (at 20°C and 1.00 atmosphere pressure) of nitrogen dioxide that theoretically should be produced by the reaction.

nitric acid (vol.)   _____

NO_2 (mass)   _____

NO_2 (volume)   _____

APPENDICES

Appendix A

SOLUTIONS TO STARRED* PRELABORATORY EXERCISES

Solutions to the starred problems are included in this section. Solutions to most of the remaining problems are in the *Instructor's Manual*.

Experiment 1 - INTRODUCTORY CONCEPTS, TECHNIQUES AND CHALLENGES

2. It is much easier to swirl solutions and to stopper them when they are in Erlenmeyer flasks. On the other hand it is easier to remove solids from a beaker.

4. argon, carbon, copper, gold, helium, krypton, neon, nitrogen, oxygen, platinum, radon, silver, sulfur, xenon

5.   a.   element   b.   compound   c.   mixture

7.   a.   physical   b.   chemical   c.   physical

8. The composition gives the lowest whole number ratio of the atoms of each element present in the compound (For compounds that exist in molecular form, the composition also should give the actual number of atoms of each element present in a molecule of the substance.). The structure gives the sequence of bonding and spatial arrangement of the atoms in the compound. As illustrated in the example, it is possible to have the same formula but different structures and the different structures have different properties thus they represent different compounds. Compounds with the same formula but different structures are called isomers.

Experiment 2 - MEASUREMENTS

1. a.   Units related by powers of 10 instead of 3, 12, 16 etc., distance cubed is related to volume by definition resulting in 1 cm^3 = 1 mL (how many gallons in 1 ft^3), at 3.98°C 1 mL of water weighs exactly 1 gram and it is within 0.2% at room temperture (a pint is only within 4% of a pound for water), the metric system is used internationally by most countries and it would be much better to have only one type of measurements requiring only one set of tools.

2.   a.   3   b.   2   c.   4

3. a. $(5.2 \text{ g}) \left(\dfrac{1 \text{ kg}}{10^3 \text{ g}} \right) = 5.2 \times 10^{-3} \text{ kg}$

   b. $(3.74 \times 10^{-1} \text{ km}) \left(\dfrac{10^3 \text{ m}}{1 \text{ km}} \right) \left(\dfrac{10^2 \text{ cm}}{1 \text{ m}} \right) = 3.74 \times 10^4 \text{ cm}$

   c. $(7.5 \times 10^1 \text{ cm}^2) \left(\dfrac{1 \text{ m}}{10^2 \text{ cm}} \right)^2 = 7.5 \times 10^{-3} \text{ m}^2$

       g.   9.5 cm^2 h. 2.02×10^1 g/mol   i. 9.877×10^3 cm

Experiment 2 continued

5. a. $\left(\dfrac{15 \text{ drops}}{0.60 \text{ g}}\right)\left(\dfrac{0.80 \text{ g}}{1.0 \text{ mL}}\right) \quad = \quad \dfrac{20 \text{ drops}}{1.0 \text{ mL}}$

   b. $\dfrac{1.0 \text{ mL}}{20 \text{ drops}} \quad = \quad 0.05 \text{ mL/drop}$

6. a. balance 1 average $= \dfrac{3.53 \text{ g} + 3.55 \text{ g} + 3.51 \text{ g}}{3} = 3.53 \text{ g}$

   balance 2 average $= \dfrac{3.37 \text{ g} + 3.45 \text{ g} + 3.47 \text{ g}}{3} = 3.43 \text{ g}$

   b. average deviation (bal. 1) $= \dfrac{0.00 \text{ g} + 0.02 \text{ g} + 0.02 \text{ g}}{3} = 0.01 \text{ g}$

   average deviation (bal. 2) $= \dfrac{0.06 \text{ g} + 0.02 \text{ g} + 0.04 \text{ g}}{3} = 0.04 \text{ g}$

   c. Balance 1 has the lower average deviation (0.01 g vs 0.04 g for balance 2) and is more precise.

   d. Balance 1 is more accurate as its average value is closer to the real value.

Experiment 3 - DENSITY

1. a. $d = m/v = \dfrac{108.95 \text{ g} - 89.22 \text{ g}}{25.0 \text{ mL}} = 0.789 \text{ g/mL}$

   b. $(5.00 \text{ g})\left(\dfrac{1.00 \text{ mL}}{0.789 \text{ g}}\right) \quad = \quad 6.34 \text{ mL}$

3. radius $=$ diameter$/2 = 12.5 \text{ cm}/2 = 6.25 \text{ cm}$

   $d = \dfrac{m}{v} = \dfrac{m}{\pi r^2 H} = \dfrac{1.089 \text{ g}}{\pi(6.25 \text{ cm})^2(1.62\times10^{-2} \text{ cm})} = 0.548 \text{ g/cm}^3$

8. From the graph, a density measurement of 1.030 g/cm³ indicates that the solution is 25% by mass ethylene glycol.

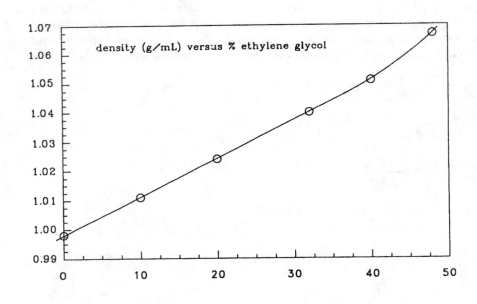

density (g/mL) versus % ethylene glycol

Experiment 4 - **MIXTURES: SEPARATION AND PURIFICATION**

3. a.   HCl is a gas and would evaporate before the water.

   b.   Calcium carbonate has an extremely low solubility in water (6×10^{-3} g/100 mL solution). Evaporation of 10 g of water would give a result indistuishable from zero on milligram balances.

   c.   Arsenic(III) chloride is highly toxic and would be hazardous to use and it decomposes in water.

4. a.   Sand is not soluble in cold or hot water.

6. a. $\dfrac{44.599 \text{ g} - 44.317 \text{ g}}{52.987 \text{ g} - 44.317 \text{ g}} \times 100\% = 3.25\%$

   b. $\dfrac{0.0325 - 0.0269}{0.0269} \times 100\% = 21\%$

Experiment 5 - **MELTING POINTS**

1. a. $\dfrac{1.8 \text{ g}}{2.5 \text{ g}} \times 100\% = 72\%$

2. No. A melting point should be taken on an intimate mixture of the unknown and cinnamic acid. If the melting point is not depressed, it is likely that the unknown is cinnamic acid. If the melting point is depressed, the unknown is neither urea nor cinnamic acid.

4. The melting point occurs at the intersection of the two lines, in this case, 70.0°C.

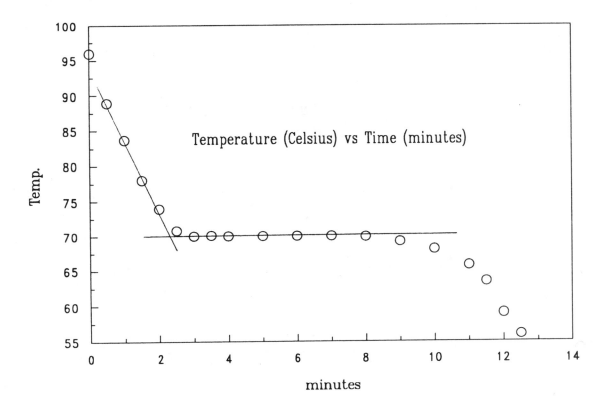

Experiment 6 - PERIODIC PROPERTIES, SOLUBILITY AND EXTRACTION

1.
   a. The latter (ethanol) should have (and does) a considerably higher boiling point than propane. Ethanol has strong intermolecular hydrogen bonds.

   d. As neither one of these compounds has a hydrogen bonded to F, O, N or Cl, neither is capable of hydrogen bonding and the predominant intermolecular attractions are due to very weak dipole-dipole and London forces. The latter are more important thus chloromethane with its greater number of electrons should and does have the higher boiling point.

2.
   a. Sodium chloride being an ionic (or extremely polar) compound should be and is much more soluble in the very polar water than in the nonpolar kerosene.

   c. Although HCl in the gas phase has polar covalent bonding, it is a very strong acid and totally dissociates in water. The resulting ions are stabilized by strong solvation by the waters. The polar HCl molecules are not attracted to the nonpolar kerosene. As a result, HCl is much more soluble in water than kerosene.

Experiment 8 - PAPER CHROMATOGRAPHY

4.

amino acid	distance (units)	R_f
aspartic acid	2.0	0.25
alanine	3.0	0.38
phenylalanine	5.0	0.62
aspartame	2.0	0.25
	5.0	0.62
leucine	5.5	0.69
valine	4.5	0.56
lysine	1.5	0.19

aspartame $\xrightarrow{\text{HCl}}$ CH_3OH + aspartic acid + phenylalanine

Experiment 9 - CLASSIFICATION OF CHEMICAL REACTIONS

1. Classify the following reactions according to combination (CA), decomposition (D), combustion (CU), single replacement (SR) or double replacement (DR) and then balance the equations.

reaction	classification
a. $\underline{1}$ Mg(s) + $\underline{1}$ ZnCl$_2$(aq) = $\underline{1}$ MgCl$_2$(aq) + $\underline{1}$ Zn(s)	__SR__
b. $\underline{2}$ AgNO$_3$(aq) + $\underline{1}$ CaCl$_2$(aq) = $\underline{2}$ AgCl(s) + $\underline{1}$ Ca(NO$_3$)$_2$(aq)	__DR__
c. $\underline{2}$ C$_2$H$_6$(g) + $\underline{7}$ O$_2$(g) = $\underline{4}$ CO$_2$(g) + $\underline{6}$ H$_2$O(g)	__CU__
d. $\underline{1}$ Na$_2$O(s) + $\underline{1}$ H$_2$O(l) = $\underline{2}$ NaOH(aq)	__CA__
e. $\underline{2}$ KClO$_3$(s) = $\underline{2}$ KCl(s) + $\underline{3}$ O$_2$(g)	__D__

2. Complete, balance and classify the following reactions:

a. For this reaction, decomposition, combustion and single replacement can be ruled out. It is hard to imagine a logical combination product so this leaves double replacement.

$\underline{1}$ BaCl$_2$(aq) + $\underline{1}$ Na$_2$SO$_4$(aq) = $\underline{1}$ BaSO$_4$(s) + $\underline{2}$ NaCl(aq)   __DR__

b. Only single replacement makes sense for these two reactants.

$\underline{2}$ Fe(s) + $\underline{3}$ CuCl$_2$(aq) = $\underline{2}$ FeCl$_3$(aq) + $\underline{3}$ Cu(s)   __SR__

c. Combination and combustion are the possible choices for this one. Although some reactions are both, this one is not a combination. Oxygen almost always supports the combustion of organic compounds to give carbon dioxide and water.

$\underline{2}$ C$_6$H$_6$(l) + $\underline{15}$ O$_2$(g) = $\underline{12}$ CO$_2$(g) + $\underline{6}$ H$_2$O(g)   __CU__

d. Combination and double replacement are the possible choices here but double replacement results in formation of a strong acid and strong base.

$\underline{1}$ BaCl$_2$(s) + $\underline{2}$ H$_2$O(l) = $\underline{1}$ Ba(OH)$_2$(aq) + $\underline{2}$ HCl(aq)

A strong acid would react with a strong base to give the reactants thus the equilibrium for this reaction lies far to the left and the reaction will not proceed left to right. This leaves combination as the only logical possibility.

$\underline{1}$ BaCl$_2$(s) + $\underline{2}$ H$_2$O(l) = $\underline{1}$ BaCl$_2$·2H$_2$O(s)   __CA__

Experiment 10 - EMPIRICAL FORMULA OF A HYDRATE

2. a. 187.55 g/mol

   c. $(2.3 \text{ g ethanol})\left(\dfrac{1 \text{ mol}}{46.0 \text{ g}}\right) \quad = \quad 5.0\times10^{-2} \text{ mol}$

   e. $(8.7\times10^{-4} \text{ mol})\left(\dfrac{249.68 \text{ g}}{1 \text{ mol}}\right) \quad = \quad 2.17\times10^{-1} \text{ g}$

   g. $\text{C} \quad \dfrac{36.03}{60.10} \times 100\% = 59.95\% \qquad\qquad \text{H} \quad \dfrac{8.062}{60.10} \times 100\% = 13.41\%$

   $\text{O} \quad \dfrac{16.00}{60.10} \times 100\% = 26.62\%$

   i. $\dfrac{2 \times 18.016 \text{ g/mol}}{147.02 \text{ g/mol}} \times 100\% = 24.51\%$

3. Assume 100.0 g of freon 11.

   $(8.74 \text{ g C})\left(\dfrac{1 \text{ mol C}}{12.011 \text{ g C}}\right) \quad = \quad 0.728 \text{ mol C} \qquad 0.728/0.728 = 1$

   $(77.43 \text{ g Cl})\left(\dfrac{1 \text{ mol Cl}}{35.453 \text{ g Cl}}\right) \quad = \quad 2.184 \text{ mol Cl} \qquad 2.184/0.728 = 3 \qquad \text{CFCl}_3$

   $(13.83 \text{ g F})\left(\dfrac{1 \text{ mol F}}{18.998 \text{ g F}}\right) \quad = \quad 0.7280 \text{ mol F} \qquad 0.7280/0.728 = 1$

6. $\dfrac{0.793 \text{ g}}{4.00 \text{ g}} \times 100\% = 19.8\% \text{ water}$

   $3.21 \text{ g NiBr}_2 \times \dfrac{1 \text{ mol NiBr}_2}{218.5 \text{ g NiBr}_2} = 0.0147 \text{ moles NiBr}_2$

   $0.793 \text{ g H}_2\text{O} \times \dfrac{1 \text{ mole H}_2\text{O}}{18.0 \text{ g H}_2\text{O}} = 0.441 \text{ moles H}_2\text{O}$

   $\dfrac{0.0441 \text{ moles H}_2\text{O}}{0.0147 \text{ mol NiBr}_2} = 3.00 \text{ moles H}_2\text{O/mol NiBr}_2$ **Therefore the formula is $\text{NiBr}_2{\cdot}3\text{H}_2\text{O}$**

Experiment 11 - STOICHIOMETRY OF A REACTION

1. b. 295.64 g/mol

   d. $(3.6\times10^{-2} \text{ g})\left(\dfrac{1 \text{ mol}}{184.24 \text{ g}}\right) \quad = \quad 2.0\times10^{-4} \text{ mol}$

   f. $(4.5 \text{ mol})\left(\dfrac{404.00 \text{ g}}{1 \text{ mol}}\right) = 1.8\times10^{3} \text{ g}$

2. $(2.7 \text{ g Al})\left(\dfrac{1 \text{ mol Al}}{27.0 \text{ g Al}}\right)\left(\dfrac{3 \text{ mol H}_2}{2 \text{ mol Al}}\right)\left(\dfrac{2.02 \text{ g H}_2}{1 \text{ mol H}_2}\right) = 0.30 \text{ g H}_2$

3. a. Theoretical yield:

   $(69 \text{ g SA})\left(\dfrac{1 \text{ mol SA}}{138 \text{ g SA}}\right)\left(\dfrac{1 \text{ mol Asp}}{1 \text{ mol SA}}\right)\left(\dfrac{180 \text{ g Asp}}{1 \text{ mol Asp}}\right) = 9.0\times10^{1} \text{ g}$

   Percent yield:

   $\dfrac{72 \text{ g Asp}}{9.0\times10^{1} \text{ g Asp}} \times 100\% = 80\%$

Experiment 12 - ENTHALPIES IN PHYSICAL AND CHEMICAL CHANGES

1. a.   +   b.   −

2. $(m_e \times \Delta t_d \times C)$ + $(\Delta H_f \times m_s)$ + $(m_s \times \Delta t_{in} \times C)$ = 0
    energy lost by liquid   energy used to melt   energy used to raise
    ethylene glycol   ethylene glycol   temperature of melted
                                                                      ethylene glycol

m_e =   mass of liquid ethylene glycol
Δt_d =   temperature drop of liquid ethylene glycol
C =   specific heat of liquid ethylene glycol   =   2.3 J/g-K
ΔH_f =   heat of fusion of ethylene glycol
m_s =   mass of solid ethylene glycol
Δt_{in} =   temperature increase of melted ethylene glycol

$(1.00 \times 10^2)(-14.0)(2.3)$ + $(\Delta H_f)(15)$ + $(15)(15.5)(2.3)$ =   0

$$\Delta H_f \quad = \quad 1.8 \times 10^2 \text{ J/g}$$

4.
   a. $HNO_3(aq)$ + $KOH(aq)$ = $KNO_3(aq)$ + $H_2O(l)$

        H^+ + OH^- = $H_2O(aq)$

   b. $2 HCl(aq)$ + $Sr(OH)_2(aq)$ = $SrCl_2(aq)$ + $2 H_2O(l)$

        H^+ + OH^- = $H_2O(aq)$

5. The limiting reagent is sodium hydroxide. One mole of sodium hydroxide will completely react with 0.5 mole of H_2SO_4.

6. a.   trinitrotoluene (TNT), nitro glycerine, ammonium nitrate

---

Experiment 13 - CHEMICAL PROPERTIES OF OXYGEN AND HYDROGEN

1.
   a. $N_2(g)$ + $3 H_2(g)$ = $2 NH_3(g)$

   b. $\dfrac{1 \text{ mol } N_2}{3 \text{ mol } H_2}$

5.
   a.   The gas must have a low solubility in water and must not react with water.

Experiment 14 - GAS LAWS

1. $0.90 \text{ atm} \times \dfrac{760 \text{ mm}}{1 \text{ atm}} = 680 \text{ mm}$

2. $V_2 = V_1 \times \dfrac{P_1}{P_2} = 75 \text{ mL} \times \dfrac{730 \text{ mm}}{780 \text{ mm}} = 7.0 \times 10^1 \text{ mL}$

5. $P = 735 \text{ mm} \times \dfrac{1 \text{ atm}}{760 \text{ mm}} = 0.967 \text{ atm} \qquad T = 273 + 25 = 298 \text{ K}$

    $PV = nRT \qquad n = PV/RT$

    $n = \dfrac{(0.967 \text{ atm})(0.548 \text{ L})}{(0.08206 \text{ L-atm/mol-K})(298 \text{ K})} = 0.0217 \text{ mole}$

    $\dfrac{0.500 \text{ g}}{0.0217 \text{ mol}} = 23.0 \text{ g/mol} \qquad \dfrac{24.31 \text{ g/mol} - 23.0 \text{ g/mol}}{24.31 \text{ g/mol}} \times 100\% = 5.4\% \text{ error}$

6. $750 \text{ mm} \times \dfrac{1 \text{ atm}}{760 \text{ mm}} = 0.987 \text{ atm} \qquad T = 273 + 27 = 300 \text{ K}$

    $d = \dfrac{1.32 \text{ g}}{0.750 \text{ L}} = 1.76 \text{ g/L}$

    $M = \dfrac{dRT}{P} = \dfrac{(1.76 \text{ g/L})(0.08206 \text{ L-atm/mol-K})(300 \text{ K})}{0.987 \text{ atm}} = 43.9 \text{ g/mol}$

Experiment 15 - DISTILLATION AND HARDNESS OF H_2O

5.
   a. 95.2°C

   b. The average energy of the molecules at 95.2°C is lower than at 100°C and the reaction or cooking rate will be slower.

   c. The pressure in a pressure cooker builds up above 1 atmosphere which raises the boiling point and consequently increases the cooking rate.

6. The yellow flame test indicates the presence of sodium and the positive silver nitrate test indicates the probable presence of chloride or another anion which forms an insoluble compound with silver.

7. $\dfrac{(500)(6.25 \text{ mL EDTA})}{25.00 \text{ mL } H_2O} = 125 \text{ ppm}$

Experiment 16 - IONIC REACTIONS AND CONDUCTIVITY

1. KCl, HNO_3, KOH and Na_2SO_4 are electrolytes. All are soluble in water and dissociate into ions when they dissolve.

   Although $AgCl$ and $CaCO_3$ are ionic, they have extremely low solubilities in water and do not provide the solution with enough ions to conduct a current.

   Isopropyl alcohol and acetone are both soluble in water but neither one ionizes and they are both nonelectrolytes.

2.
   a. FE $Ba(OH)_2(aq) + Na_2CO_3(aq) = BaCO_3(s) + 2\,NaOH(aq)$
                                                   white ppt.

       TIE $Ba^{2+} + 2\,OH^- + 2\,Na^{2+} + CO_3^{2-} = BaCO_3(s) + 2\,Na^+ + 2OH^-$

       NIE $Ba^{2+} + CO_3^{2-} = BaCO_3(s)$

   b. FE $Ba(OH)_2(aq) + 2\,HCl(aq) = BaCl_2(aq) + 2\,H_2O(l) + \textit{heat}$

       TIE $Ba^{2+} + 2\,OH^- + 2\,H^+ + 2\,Cl^- = Ba^{2+} + 2\,Cl^- + 2\,H_2O(l)$

       NIE $H^+ + OH^- = H_2O(l)$

   c. FE $Na_2CO_3(aq) + 2\,HCl(aq) = 2\,NaCl(aq) + H_2O(aq) + CO_2(g)$
                                                                    gas

       TIE $2\,Na^+ + CO_3^{2-} + 2\,H^+ + 2\,Cl^- = 2\,Na^+ + 2\,Cl^- + H_2O(l) + CO_2(g)$

       NIE $2\,H^+ + CO_3^{2-} = H_2O(l) + CO_2(g)$

   A = HCl B = $Ba(OH)_2$ C = Na_2CO_3

═══

Experiment 17 - ANALYSIS OF CATIONS

1. $Pb^{2+} + 2\,Cl^- = PbCl_2(s)$

    $Hg_2^{2+} + 2\,Cl^- = Hg_2Cl_2(s)$

    $Ag^+ + Cl^- = AgCl(s)$

2. $PbCl_2(s) = Pb^{2+} + 2\,Cl^-$

3. $Pb^{2+} + CrO_4^{2-} = PbCrO_4(s)$

4. $Hg_2Cl_2(s) + 2\,NH_3(aq) = Hg(l) + HgNH_2Cl(s) + NH_4^+ + Cl^-$

    $AgCl(s) + 2\,NH_3(aq) = Ag(NH_3)_2^+ + Cl^-$

5. $Ag(NH_3)_2^+ + 2\,H^+ + Cl^- = AgCl(s) + 2\,NH_4^+$

Experiment 17 continued

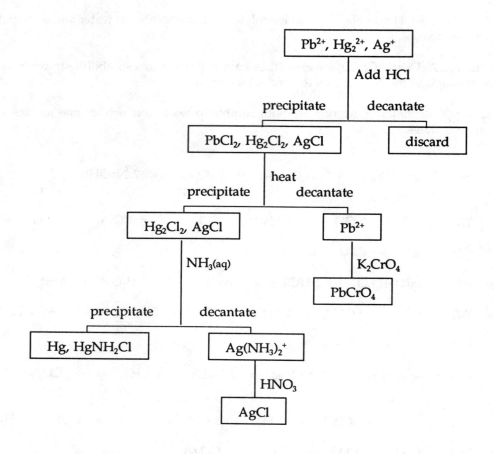

Experiment 18 - SPECTROSCOPY

1. $\left(\dfrac{2.18 \text{ g Co(NO}_3)_2 \cdot 6\text{H}_2\text{O}}{0.0500 \text{ L}} \right) \left(\dfrac{1 \text{ mol Co(NO}_3)_2 \cdot 6\text{H}_2\text{O}}{291.04 \text{ g Co(NO}_3)_2 \cdot 6\text{H}_2\text{O}} \right) \; = \; 0.150 \text{ moles/L}$

6. concentration of stock solution $= \dfrac{2.47 \times 10^{-3} \text{ mol}}{0.250 \text{ L}} = 9.88 \times 10^{-3} \text{ M}$

sample calculation for concentration of iron(III) salicylate in tube #3:

$M_1 V_1 = M_2 V_2$

$M_2 = \dfrac{M_1 V_1}{V_2} = \dfrac{(9.88 \times 10^{-3} \text{ mol/L})(3 \text{ mL})}{100 \text{ mL}} = 2.96 \times 10^{-4} \text{ M}$

Experiment 18 continued

solution #	Concentration (mol/L)	Absorption
1	9.88×10^{-5}	0.17
2	1.98×10^{-4}	0.33
3	2.96×10^{-4}	0.51
4	3.95×10^{-4}	0.68
5	4.94×10^{-4}	0.86
unknown	3.6×10^{-4}	0.62

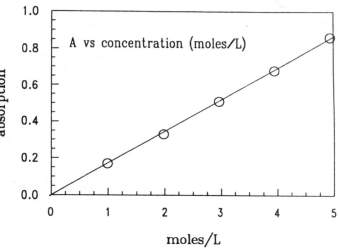

Experiment 19 - ACIDITY AND pH

1.

pH	$[H^+]$ (moles/L)	pOH	$[OH^-]$ (moles/L)
4	1×10^{-4}	10	1×10^{-10}
6	1×10^{-6}	8	1×10^{-8}
11	1×10^{-11}	3	1×10^{-3}
9	1×10^{-9}	5	1×10^{-9}
1.44	3.6×10^{-2}	12.56	2.8×10^{-13}
8.14	7.2×10^{-9}	5.86	1.4×10^{-6}

3. HNO_3 is a strong acid and ionizes almost completely according to: $HNO_{3(aq)} = H^+ + NO_3^-$ The H^+ concentration in a 1×10^{-2} M HNO_3 solution would be 1×10^{-2} M and the pH would be 2.

5. H_2S ionizes partially according to $H_2S_{(aq)} = H^+ + SH^-$. Every time an H^+ ion is formed, an SH^- ion is formed. As the pH of the solution was 4, the $[H^+] = [SH^-] = 1 \times 10^{-4}$ M. The H_2S concentration would be decreased from its original concentration by the amount of ionization or the amount of hydrogen ion formed: $1.0 \times 10^{-1} - 1 \times 10^{-4} = 1.0 \times 10^{-1}$ M or for all practical purposes, its concentration is not significantly changed by the ionization.

$$K_a = \frac{[H^+][SH^-]}{[H_2S]} = \frac{(1 \times 10^{-4})(1 \times 10^{-4})}{1.0 \times 10^{-1}} = 1 \times 10^{-7}$$

6. This solution contains a weak acid (HF) and its conjugate base (F^-) and is therefore a buffer solution. HF dissociates according to : $HF_{(aq)} = H^+ + F^-$ with $K_a = \frac{[H^+][F^-]}{[HF]}$. The concentrations of HF

and F^- will not change significantly from their original values when equilibrium is attained therefore:

$H^+ = $ the antilogarithm of $-3.45 = 3.5 \times 10^{-4}$ M, $K_a = \frac{(3.5 \times 10^{-4})(0.1)}{0.1} = 3.5 \times 10^{-4}$

7. $\quad C_2H_3O_2^- + H_2O_{(aq)} = HC_2H_3O_{2(aq)} + OH^-$

Experiment 20 - SYNTHESIS

1. Theoretical yield of salicylic acid (MS = methylsalicylate, SA = salicylic acid)

$$(25 \text{ g MS}) \left(\frac{1 \text{ mol MS}}{152 \text{ g MS}} \right) \left(\frac{1 \text{ mol SA}}{1 \text{ mol MS}} \right) \left(\frac{138 \text{ g SA}}{1 \text{ mol SA}} \right) = 23 \text{ g SA}$$

$$\text{percent yield} = \frac{\text{experimental yield}}{\text{theoretical yield}} \times 100\% = \frac{16 \text{ g}}{23 \text{ g}} \times 100\% = 70\%$$

2. Theoretical yield of aspirin: (SA = salicylic acid, A = aspirin)

$$(16 \text{ g SA}) \left(\frac{1 \text{ mol SA}}{138 \text{ g SA}} \right) \left(\frac{1 \text{ mol A}}{1 \text{ mol SA}} \right) \left(\frac{180 \text{ g A}}{1 \text{ mol A}} \right) = 21 \text{ g A}$$

$$\text{percent yield} = \frac{16 \text{ g}}{21 \text{ g}} \times 100\% = 76\%$$

Experiment 21 - ACID-BASE TITRATIONS

1. KHP = potassium hydrogen phthalate

$$\left(\frac{0.6530 \text{ g KHP}}{23.32 \text{ mL NaOH}} \right) \left(\frac{1 \text{ mol KHP}}{204.23 \text{ g KHP}} \right) \left(\frac{1 \text{ mol NaOH}}{1 \text{ mol KHP}} \right) \left(\frac{10^3 \text{ mL NaOH}}{1 \text{ L NaOH}} \right) = 0.1371 \text{ mol NaOH/L}$$

2. $$\left(\frac{0.0160 \text{ L NaOH}}{0.0250 \text{ L HCl}} \right) \left(\frac{0.120 \text{ mol NaOH}}{1 \text{ L NaOH}} \right) \left(\frac{1 \text{ mol HCl}}{1 \text{ mol NaOH}} \right) = 0.0768 \text{ mol HCl/L}$$

3. $$(0.0213 \text{ L NaOH}) \left(\frac{0.120 \text{ mol NaOH}}{1 \text{ L NaOH}} \right) \left(\frac{1 \text{ mol HA}}{1 \text{ mol NaOH}} \right) = 2.56 \times 10^{-3} \text{ moles HA}$$

$$\frac{0.400 \text{ g}}{2.56 \times 10^{-3} \text{ mol HA}} = 156 \text{ g HA/mol}$$

7. The products of the reaction are sodium chloride and water. As sodium chloride should not affect the pH of the solution, the indicator should ideally change at a pH of 7.

Experiment 22 - OXIDATION - REDUCTION

1. Zinc. Zinc is oxidized in the reaction and reduces the nickel. If the reaction did not go, then nickel would have been the better reducing agent.

oxidation half reaction $\qquad Zn_{(s)} = Zn^{2+} + 2 e^-$

reduction half reaction $\qquad 2 e^- + Ni^{2+} = Ni_{(s)}$

net ionic reaction $\qquad Zn_{(s)} + Ni^{2+} = Zn^{2+} + Ni_{(s)}$

2. Chlorine. The observation that the reaction does not go implies that the reverse reaction would go spontaneously. In the reverse reaction, Cl_2 oxidizes the Pb and is therefore a better oxidizing agent than Pb^{2+}.

Experiment 23 - ANALYSIS OF BLEACH AND COPPER(II) GLYCINATE

1. Pool "chlorine"

   Molarity of NaClO

   $$\left(\frac{21.12 \text{ mL Na}_2\text{S}_2\text{O}_3}{1.00\times10^{-3} \text{ L NaClO}}\right)\left(\frac{1 \text{ L Na}_2\text{S}_2\text{O}_3}{10^3 \text{ mL Na}_2\text{S}_2\text{O}_3}\right)\left(\frac{0.150 \text{ mol Na}_2\text{S}_2\text{O}_3}{1 \text{ L Na}_2\text{S}_2\text{O}_3}\right)\left(\frac{1 \text{ mol NaClO}}{2 \text{ mol Na}_2\text{S}_2\text{O}_3}\right) = 1.58 \text{ M NaClO}$$

   $$\left(\frac{1.58 \text{ mol NaClO}}{1 \text{ L}}\right)\left(\frac{74.44 \text{ g NaClO}}{1 \text{ mol NaClO}}\right)\left(\frac{1 \text{ L}}{1.18\times10^3 \text{ g}}\right)(100\%) = 10.0\%$$

2. Mass percent copper in copper(II) acetate monohydrate

   Experimental %

   $$\left(\frac{1.036\times10^{-2} \text{ L}}{0.250 \text{ g sample}}\right)\left(\frac{0.1200 \text{ moles Na}_2\text{S}_2\text{O}_3}{1 \text{ L Na}_2\text{S}_2\text{O}_3}\right)\left(\frac{1 \text{ mol Cu}^{2+}}{1 \text{ mol Na}_2\text{S}_2\text{O}_3}\right)\left(\frac{63.54 \text{ g Cu}}{1 \text{ mol Cu}}\right)(100\%) = 31.6\%$$

   Theoretical %

   $$\left(\frac{63.54}{199.65}\right)(100\%) = 31.83\%$$

Experiment 24 - THE RATES OF CHEMICAL REACTIONS

1.

run #	1	2	3	4	5	6	7
1/t (1/sec.)	0.025	0.013	0.0061	0.026	0.024	0.012	0.0063

The rate is proportional to the concentrations of A and C but independent of the concentration of B. The rate expression then is:

$$\frac{\Delta[P]}{\Delta t} = k[A][C]$$

This rate expression suggests a two step mechanism. The first step would be the rate determining or slow step, A + C = I (intermediate) followed by a fast step, I + B = P.

A + C = I   slow

I + B = P   fast

4. Solution 2 functions as the timer for the reaction. When the thiosulfate is depleted, the iodine produced by the reaction will not disappear and complexes with the starch to form the purple color.

5. Solutions 3 and 4 are added to replace solutions 1 and 6 to maintain the same ionic strength of the solution.

Experiment 25 - EQUILIBRIUM STUDIES

1. a. $$K_{eq} = \frac{[PCl_3][Cl_2]}{[PCl_5]}$$

2.

a. $K_{sp} = [Ca^{2+}][SO_4^{2-}]$

b. The addition of calcium ion to the solution should cause the system to shift to the left resulting in the precipitation of calcium sulfate.

c. The addition of sulfate ion to the solution should cause the system to shift to the left resulting in the precipitation of calcium sulfate.

d. Neither sodium ions or chloride ions are involved in the dissolving of calcium sulfate and the addition of sodium chloride should not cause an observable result.

COMMON IONS BY CHARGE

A. Positive Ions

+ 1

Ammonium	NH_4^+	Mercury(I) (mercurous)	Hg_2^{2+}
Copper(I) (cuprous)	Cu^+	Potassium	K^+
Hydrogen	H^+	Silver	Ag^+
Hydronium	H_3O^+	Sodium	Na^+
Lithium	Li^+		

+ 2

Barium	Ba^{2+}	Magnesium	Mg^{2+}
Cadmium	Cd^{2+}	Manganese(II) (manganous)	Mn^{2+}
Calcium	Ca^{2+}	Mercury(II) (mercuric)	Hg^{2+}
Cobalt(II) (cobaltous)	Co^{2+}	Nickel(II) (nickelous)	Ni^{2+}
Copper(II) (cupric)	Cu^{2+}	Strontium	Sr^{2+}
Iron(II) (ferrous)	Fe^{2+}	Tin(II) (stannous)	Sn^{2+}
Lead(II) (plumbous)	Pb^{2+}	Zinc	Zn^{2+}

+ 3

Aluminum	Al^{3+}	Cerium(III) (cerous)	Ce^{3+}
Antimony(III)	Sb^{3+}	Chromium(III) (chromic)	Cr^{3+}
Arsenic(III)	As^{3+}	Iron(III) (ferric)	Fe^{3+}
Bismuth(III)	Bi^{3+}		

+ 4

Lead(IV) (plumbic)	Pb^{4+}	Tin(IV) (stannic)	Sn^{4+}

+ 5

Antimony(V)	Sb^{5+}	Bismuth(V)	Bi^{5+}
Arsenic(V)	As^{5+}		

COMMON IONS BY CHARGE continued

B. Negative ions

-1

Acetate	$C_2H_3O_2^-$	Hydrogen sulfate (bisulfate)	HSO_4^-
Bromate	BrO_3^-	Hydrogen sulfite (bisulfite)	HSO_3^-
Bromide	Br^-	Hydroxide	OH^-
Chlorate	ClO_3^-	Hypochlorite	ClO^-
Chloride	Cl^-	Iodate	IO_3^-
Chlorite	ClO_2^-	Iodide	I^-
Cyanate	NCO^-	Nitrate	NO_3^-
Cyanide	CN^-	Nitrite	NO_2^-
Fluoride	F^-	Perchlorate	ClO_4^-
Hydride	H^-	Permanganate	MnO_4^-
Hydrogen carbonate (bicarbonate)	HCO_3^-	Thiocyanate	SCN^-

-2

Carbonate	CO_3^{2-}	Sulfate	SO_4^{2-}
Chromate	CrO_4^{2-}	Sulfide	S^{2-}
Dichromate	$Cr_2O_7^{2-}$	Sulfite	SO_3^{2-}
Oxalate	$C_2O_4^{2-}$	Tetrathionate	$S_4O_6^{2-}$
Oxide	O^{2-}	Thiosulfate	$S_2O_3^{2-}$
Persulfate	$S_2O_8^{2-}$		

-3

Ferricyanide	$Fe(CN)_6^{3-}$	Phosphate	PO_4^{3-}

-4

Ferrocyanide	$Fe(CN)_6^{4-}$

Appendix C

SOLUBILITIES OF IONIC COMPOUNDS - APPROXIMATE # OF GRAMS OF SOLUTE PER 100 GRAMS OF SOLUTION

	$C_2H_3O_2^-$	Br^-	CO_3^{2-}	Cl^-	CrO_4^{2-}	$Fe(CN)_6^{4-}$	$Fe(CN)_6^{3-}$	OH^-	IO_3^-	I^-	NO_3^-	$C_2O_4^{2-}$	PO_4^{3-}	SO_4^{2-}	S^{2-}	SCN^-
Al^{3+}	ss	s		31			ss	1×10^{-4}		s,d	42	i	i	27	d	
NH_4^+	60	43	50	27	25	vs	s	47	2	63	66	4	26	43	vs	63
Ba^{2+}	42	51	2×10^{-3}	26	4×10^{-4}		.1	4	.02	68	8	1×10^{-2}	i	2×10^{-4}	d	26
Ca^{2+}	26	59	6×10^{-3}	43	14		36	.16	.3	68	56	7×10^{-4}	2×10^{-3}	.2	.02	s
Ce^{3+}	20	3	i	50			i	i	.1	s	64	4×10^{-5}	i	9	i	
Co^{2+}	s	54	i	35	i	i	i	3×10^{-4}	1	65	50	3×10^{-3}	i	26	4×10^{-4}	51
Cu^{2+}	7	56	i	42	i	i	i	3×10^{-4}	.1	1.1	55	2×10^{-3}	i	17	2×10^{-4}	d
Fe^{3+}		s		70		i	i	1×10^{-5}	.04		46		i	ss	3×10^{-17}	vs
Pb^{2+}	31	.8	1×10^{-4}	1	7×10^{-6}	ss	i	.02	2×10^{-3}	.07	35	1×10^{-4}	1×10^{-5}	4×10^{-3}	9×10^{-5}	.05
Li^+	75	62	1.3	45	50			11.3	45	62	43	7	.03	26	vs	vs
Mg^{2+}	40	50	.07	35	42		25	2×10^{-3}	8	58	41	.03	.02	28	d	
K^+	70	40	52	25	39	31	23	53	7.5	59	27	26	47	10	s	67
Ag^+	1.0	8×10^{-6}	3×10^{-3}	2×10^{-4}	4×10^{-3}	7×10^{-5}	i		4×10^{-3}	3×10^{-6}	70	3×10^{-3}	6×10^{-4}	.8	7×10^{-13}	2×10^{-5}
Na^+	32	48	22	26.4	47	23	15	52	8	64	47	3.3	11	20	16	58
Sr^{2+}	27	50	1×10^{-3}	35	.12		33	1	.03	64	42	5×10^{-3}	i	.01	s,d	vs
Zn^{2+}	25	82	2×10^{-2}	79	i	i	i	4×10^{-4}	.9	83	56	7×10^{-4}	i	30	10^{-8}	s

An arbitrary standard for solubility is that a compound is called soluble if at least 1 gram dissolves in 100 mL of solution.

When quantitative data could not be located, the symbols below were used:

vs = very soluble, s = soluble, ss = slightly soluble, i = insoluble, d = decomposes

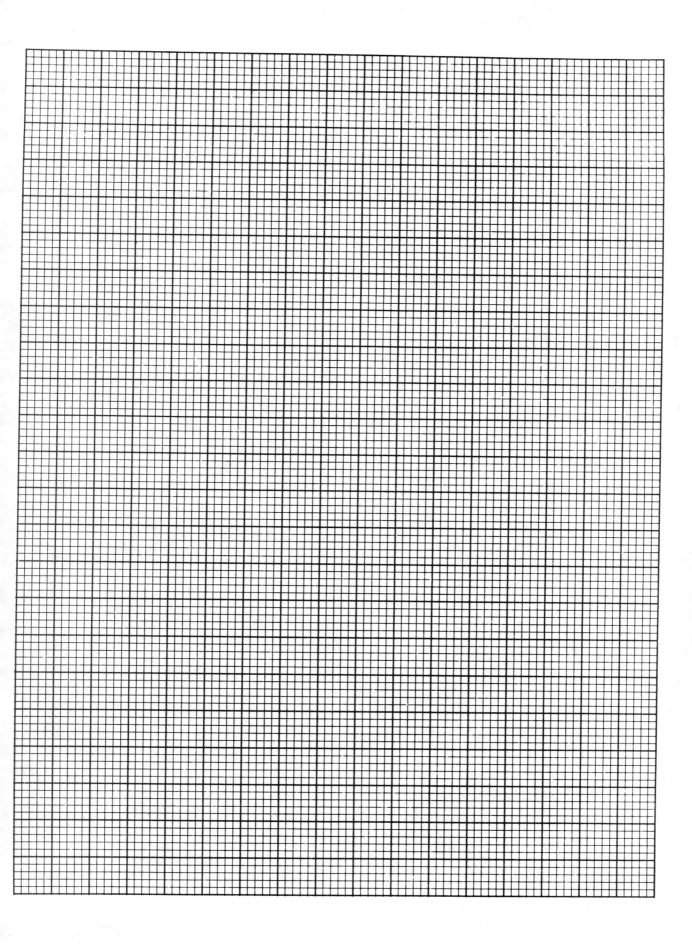

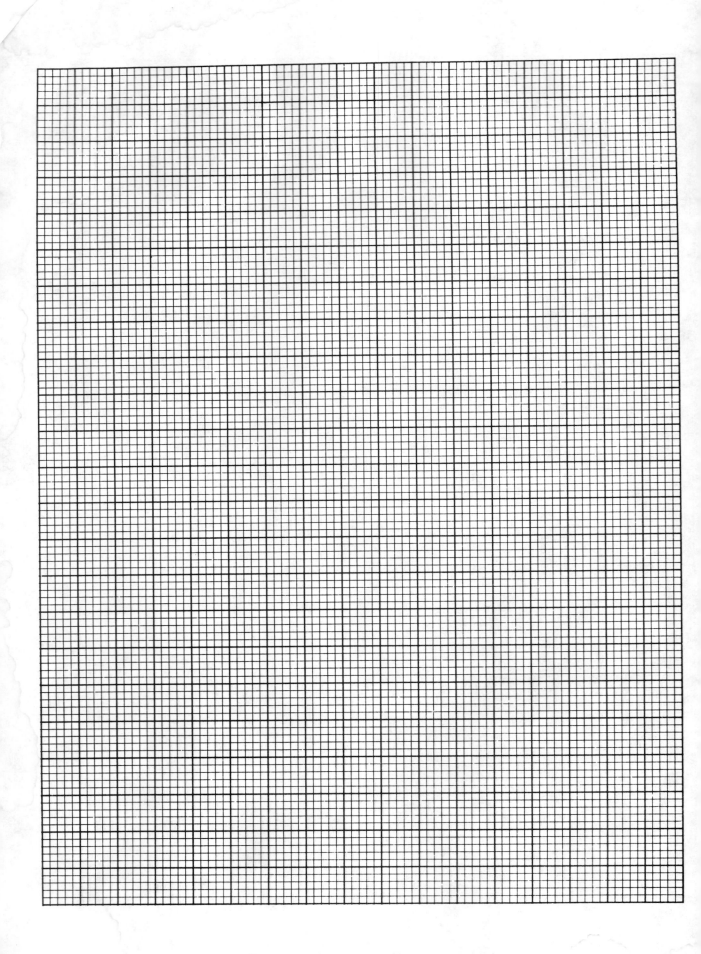

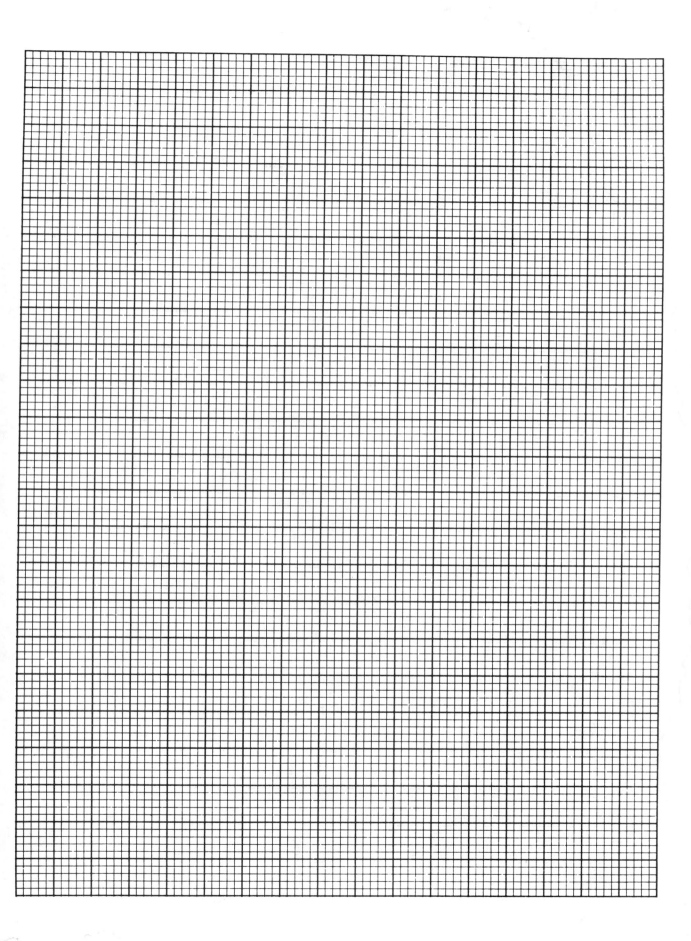

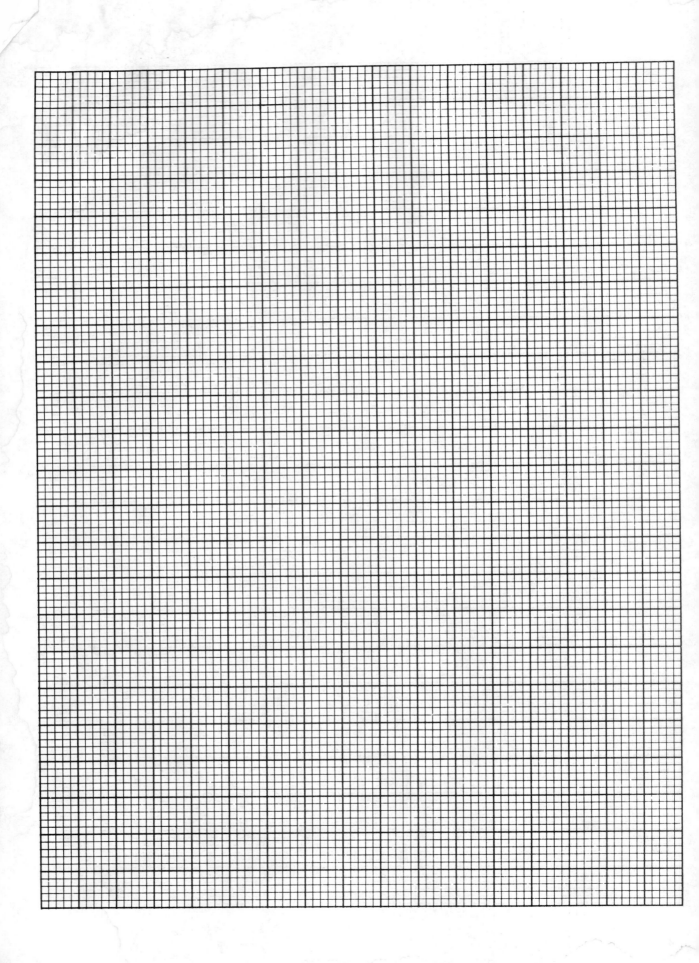

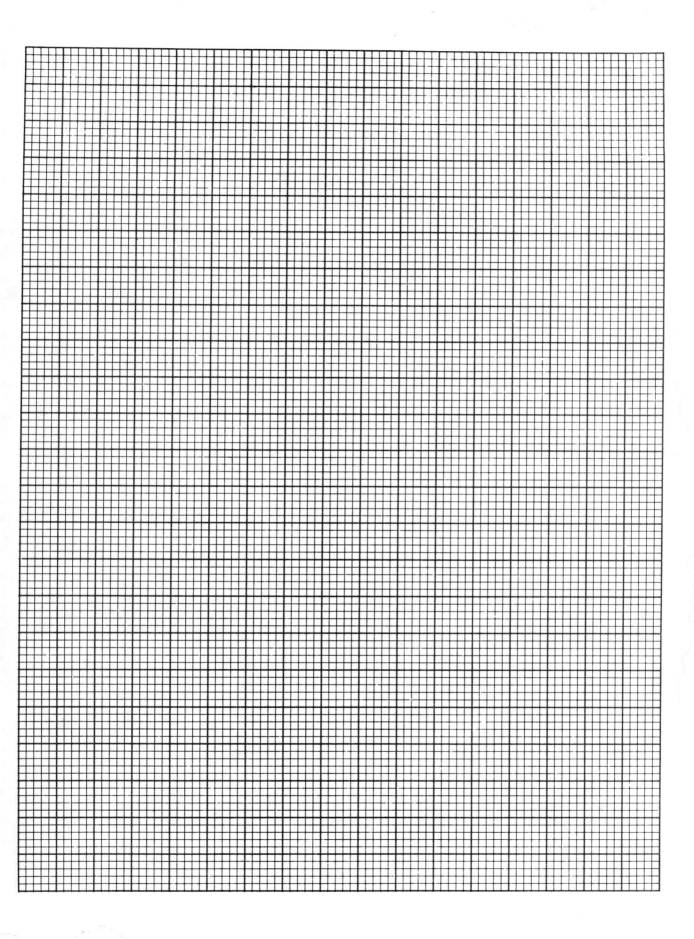

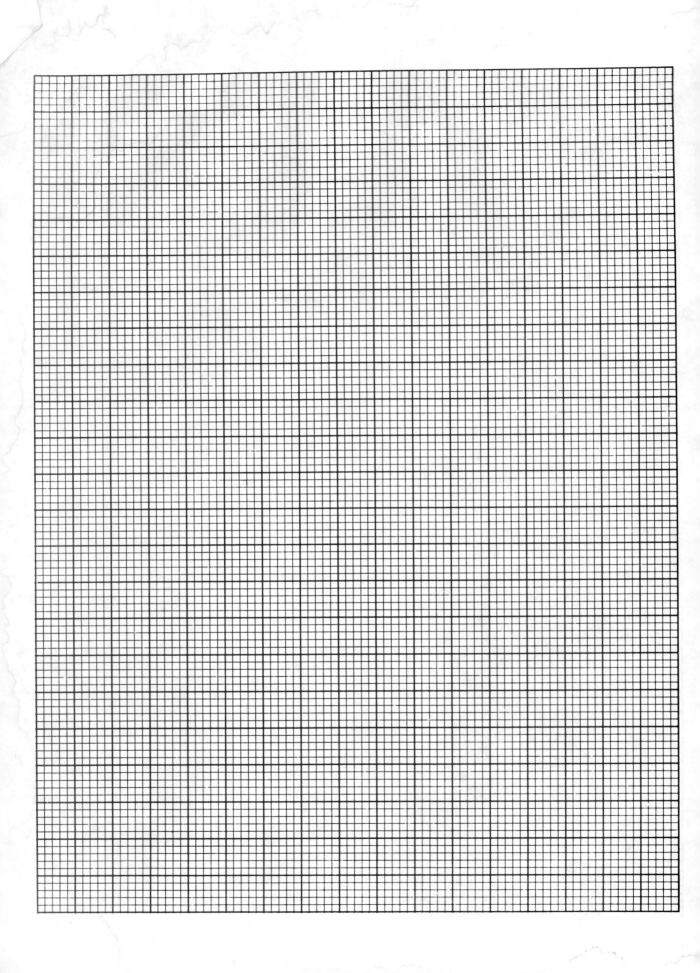

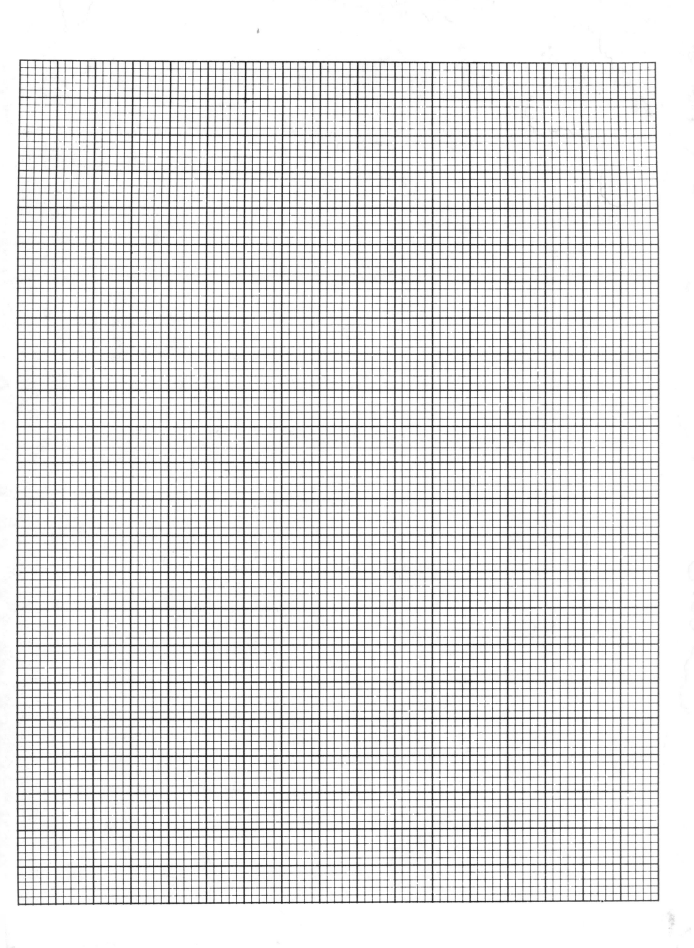

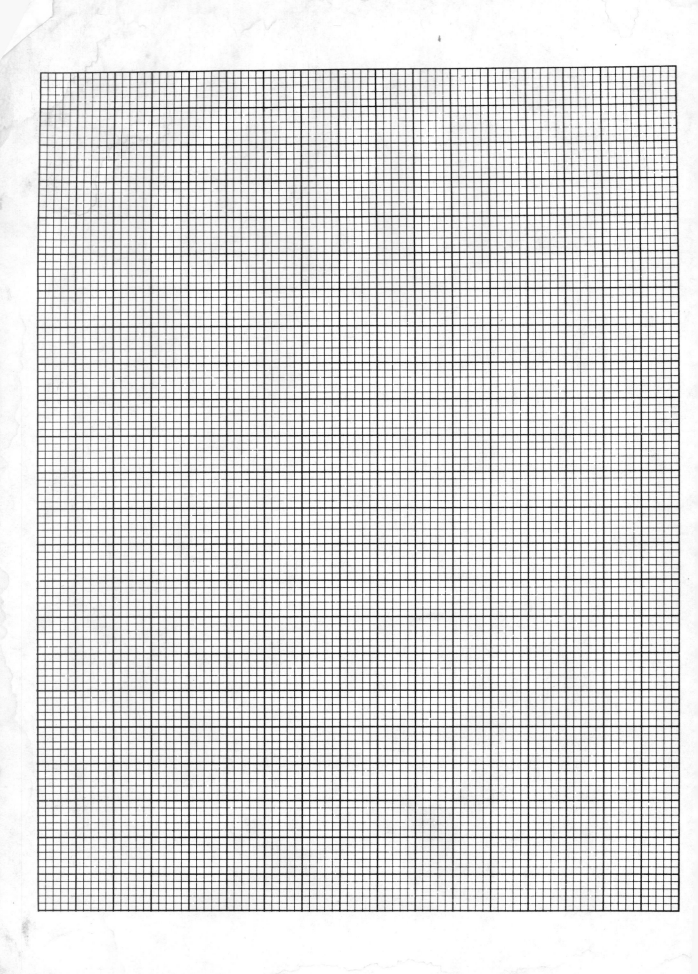